Springer Theses

Recognizing Outstanding Ph.D. Research

Aims and Scope

The series "Springer Theses" brings together a selection of the very best Ph.D. theses from around the world and across the physical sciences. Nominated and endorsed by two recognized specialists, each published volume has been selected for its scientific excellence and the high impact of its contents for the pertinent field of research. For greater accessibility to non-specialists, the published versions include an extended introduction, as well as a foreword by the student's supervisor explaining the special relevance of the work for the field. As a whole, the series will provide a valuable resource both for newcomers to the research fields described, and for other scientists seeking detailed background information on special questions. Finally, it provides an accredited documentation of the valuable contributions made by today's younger generation of scientists.

Theses are accepted into the series by invited nomination only and must fulfill all of the following criteria

- They must be written in good English.
- The topic should fall within the confines of Chemistry, Physics, Earth Sciences, Engineering and related interdisciplinary fields such as Materials, Nanoscience, Chemical Engineering, Complex Systems and Biophysics.
- The work reported in the thesis must represent a significant scientific advance.
- If the thesis includes previously published material, permission to reproduce this must be gained from the respective copyright holder.
- They must have been examined and passed during the 12 months prior to nomination.
- Each thesis should include a foreword by the supervisor outlining the significance of its content.
- The theses should have a clearly defined structure including an introduction accessible to scientists not expert in that particular field.

More information about this series at http://www.springer.com/series/8790

Eduardo Mendive Tapia

Ab initio Theory of Magnetic Ordering

Electronic Origin of Pair- and Multi-Spin Interactions

Doctoral Thesis accepted by
University of Warwick, Coventry, UK

 Springer

Author
Dr. Eduardo Mendive Tapia
Department of Computational
Materials Design
Max Planck Institute for Iron Research
Düsseldorf, Germany

Supervisor
Prof. Julie B. Staunton
Department of Physics
University of Warwick
Coventry, UK

ISSN 2190-5053   ISSN 2190-5061   (electronic)
Springer Theses
ISBN 978-3-030-37240-8   ISBN 978-3-030-37238-5   (eBook)
https://doi.org/10.1007/978-3-030-37238-5

This Springer imprint is published by the registered company Springer Nature Switzerland AG
The registered company address is: Gewerbestrasse 11, 6330 Cham, Switzerland

Supervisor's Foreword

Accurate models of magnetic interactions are needed to adapt materials for many applications and this thesis describes a major advance in this area. Discovery and exploitation of varied magnetic structures or magnetic phases, can lead to new or optimised device functionality as well as being of profound fundamental interest in condensed matter physics. For example, the exploitation of effects around magnetic phase transitions has emerged as a promising way for a new and environmentally friendly solid-state cooling technology. Refrigeration using magnetic materials is much more energy efficient than conventional cooling technologies and spans a broad temperature range around 0°C. Randomly oriented magnetic moments align when an external stimulus, such as a magnetic field is applied making the solid warm up. By removing this heat using a heat transfer fluid, like water or air, and then removing the stimulus allows the magnetic material to lower its temperature. The heat from the object being cooled is then extracted with the heat transfer fluid and the cycle completed. The changes in entropy and temperature that happen when a field is applied describe a caloric effect. In this thesis, the theoretical and computational basis is described of a model of magnetic materials at the sub-nanoscale which can address this physics quantitatively and predictively. It is shown how the degrees of freedom (the magnetic moments or spins) of the model condense out from the septillions of interacting electrons and coordinate in a complex way.

The thesis begins with a clear accessible account for postgraduate students of the phenomenon of magnetism in condensed matter physics, spin-polarised electronic structure and the determination of effective spin interactions ab initio for further modelling of magnetic properties. It goes on to present a first principles theory based on the 'disordered local moment' (DLM)-Density Functional Theory (DFT) method to describe magnetic phase diagrams and caloric properties. In the approach, the first step is to predict potential structures among all possible stabilising phases at high temperature. The theory then describes their temperature evolution and competition, predicting first- and second-order magnetic phase transitions.

The central goal of the work is to capture how the electrons cooperate to produce local magnetic moments, and how in turn magnetic order of the moments feed back to affect the electronic interactions. The outcome is a prescription of magnetic interactions not only between pairs of moments or spins but also among groups of them, i.e. apparent multi-spin interactions. Three case studies are described comprehensively.

- Multi-spin correlations are shown to be behind much of the temperature-magnetic field long-period non-collinear magnetism of the heavy rare earth elements, Gd, Tb, Dy, Ho. They arise from how the Fermi surfaces of these metals respond to the change in magnetic order of the magnetic moments established by the lanthanide f-electrons.
- The second illustration concerns the effect of strain on the geometrically frustrated magnetic interactions in the Mn-based antiperovskite material, Mn_3GaN, and gives the design of a novel cooling cycle based on magnetoelastic coupling and multi-spin interactions. These results led to a collaboration with experimentalists from which a boost to magnetic cooling performance driven by multi-spin effects has been shown for real materials with a potential new opening for magnetic refrigeration.
- Finally, the thesis describes a tour de force by its detailed accurate modelling of the apparently frustrated magnetism of the rich Mn3A class of materials (A = Ga, Ge, Sn, Ir, Rh and Pt) in all its cubic, hexagonal and tetragonal structures. Magnetic phases and transition temperatures are produced in good agreement with experiment in all cases. For example, Mn3Pt, Mn3Rh and Mn3Ir, which all crystallise into cubic structures, have triangular antiferromagnetic states at low temperatures. Moreover, the magnetovolume origin is found as to why on cooling it is only Mn3Pt which undergoes a second-order transition from a paramagnetic to collinear antiferromagnetic state followed by a first-order transition to the triangular state.

More broadly, the thesis presentation of its ab initio theory for the Gibbs free energy of a magnetic material and associated spin-polarised electronic structure lays the ground work for much further work in computational magnetic material modelling and spintronics.

Coventry, UK   Prof. Julie B. Staunton
December 2019

Abstract

We present an *ab initio* theory to describe magnetic ordering and magnetic phase transitions at finite temperatures from pairwise and multi-spin interactions. The theory models fluctuations of local magnetic moments and the feedback from the electronic structure in response to different states of magnetic order. Its key ingredient is to assume a timescale separation between the evolution of the local moment orientations and a rapidly responsive electronic background setting them. This is the Disordered Local Moment picture grounding the theoretical framework. The method uses Density Functional Theory calculations constrained to specific local moment configurations and exploits Green's functions within a Multiple Scattering Theory to solve the Kohn-Sham equations. Two central objects are calculated as functions of magnetic order: internal magnetic fields sustaining the local moments and the lattice Fourier transform of the interactions in the paramagnetic state. We develop a methodology to extract pairwise and multi-spin interactions from the first and use the second to obtain the most potential magnetic phases. We show how the free energy of a magnetic material can be obtained from the calculation of these quantities. Our approach is able to provide thermodynamic quantities of interest, such as temperature and entropy changes, and magnetic phase diagrams for temperature, magnetic field and lattice structure. Transition temperatures and their order, as well as tri-critical points, are also obtainable. This framework allows to evaluate caloric effects for their use in refrigeration exploiting magnetism, which is a central object of study in this thesis.

We carry out major investigations for long-period magnetic phases in the heavy rare earth elements (HREs), magnetic frustration in the Mn-based antiperovskite nitride Mn_3GaN, and the diverse magnetism in Mn_3A (A = Sn, Ga, Ge, Pt, Ir, Rh). The mixing of both pairwise and higher order multi-spin interactions have been found to have profound consequences on many of these systems. We have obtained a generic HRE magnetic phase diagram which is consequent on the response of the common valence electronic structure to the f-electron magnetic moment ordering, and four-spin interactions. A model based on the lanthanide contraction to describe ferromagnetic, helical antiferromagnetic and fan phases in Gd, Tb, Dy and Ho, is presented. Our study of Mn_3GaN shows that fourth-order coupling is an important effect behind its first-order paramagnetic–antiferromagnetic triangular transition. New collinear magnetic phases are predicted at high temperatures by showing the effect of biaxial strain in this triangular magnetic phase. A very rich temperature-strain phase diagram is obtained, from which a new elastocaloric cooling cycle is designed. Finally, Mn_3A family is used as a rich example to show how to obtain potential magnetic phases and the effect of magnetovolume coupling combined with the presence of multi-spin interactions.

Publications Related to this Thesis

This thesis contains research carried out in the Theory Group of the Department of Physics between October 2014 and May 2018. The content of this thesis is my own work, unless stated otherwise, carried out under the supervision of Prof. J. B. Staunton.

Parts of this thesis have been published in the following papers:

1. Chapter 5: Pair- and four-spin interactions in the heavy rare earth elements

   - E. Mendive-Tapia and J. B. Staunton. Theory of magnetic ordering in the heavy rare earths: *Ab initio* electronic origin of pair- and four- spin interactions. *Phys. Rev. Lett.*, 118:197202, 2017.

2. Chapter 6: Frustrated magnetism in Mn-based antiperovskite Mn_3GaN

   - J. Zemen, E. Mendive-Tapia, Z. Gercsi, R. Banerjee, J. B. Staunton, and K. G. Sandeman. Frustrated Magnetism and caloric effects in Mn-based antiperovskite nitrides: *Ab initio* theory. *Phys. Rev. B*, 95:184438, 2017.

3. Chapter 7: The magnetism of Mn_3A (A = Pt, Ir, Rh, Sn, Ga, Ge)

   - E. Mendive-Tapia and J. B. Staunton. *Ab initio* theory of the Gibbs free energy and a hierarchy of local moment correlation functions in itinerant electron systems: The magnetism of the Mn_3A materials class. *Phys. Rev. B*, 99:144424, 2019.

The following publication has evolved from this doctoral dissertation:

- D. Boldrin, E. Mendive-Tapia, J. Zemen, J. B. Staunton, T. Hansen, A. Aznar, J. Ll. Tamarit, M. Barrio, P. Lloveras, J. Kim, X. Moya, and F. Cohen (2018). Multisite Exchange-Enhanced Barocaloric Response in Mn_3NiN. *Physical Review X*, 8:041035, 2018.

Acknowledgements

First, I would like to express my deepest gratitude to my supervisor Prof. Julie B. Staunton, for her invaluable help and guidance, and for teaching me so much insightful knowledge in my travel through the world of magnetism. Her supervision has been beyond this project, she has been an extraordinary example of how to be an authentic researcher at both professional and human levels.

I am also thankful to Dr. Christopher Patrick for many inspiring and fruitful discussions and for always having time to answer my numerous questions.

A special mention to the great moments with Álvaro, Anna (el Rey), Anto and Edoardo (la Reina): the funniest and most unproductive times of my PhD, and probably those where I have eaten more chocolate than ever. Siempre nos quedarán los Chupa Chups de Pamplona!

My thoughts also go to Laura, Odette and Samuele, who have been very important for me during the last years of my doctoral studies. I thank them for always being with me every time I have needed friendship and for making an inside and outside of the office. Discovering with them a new passion for climbing has been a special experience for me during these years. I can say that all the time spent with them has been incredible.

Finally, a heartfelt thank you to my family for their support and love, and especially to Sofia for being with me always. All the experiences we have lived together, although many of them far from Warwick, have been unforgettable and are a treasure in my mind.

Contents

Abbreviations

AFM	Antiferromagnetic/antiferromagnetism
ASA	Atomic Sphere Approximation
bcc	Body centred cubic
BCE	Barocaloric Effect
BZ	Brillouin zone
CPA	Coherent Potential Approximation
DFT	Density Functional Theory
dis-HAFM	Distorted helical antiferromagnetic/antiferromagnetism
DLM	Disordered Local Moment (theory)
DOS	Density of states
eCE	Elastocaloric Effect
ECE	Electrocaloric Effect
FIM	Ferrimagnetic/ferrimagnetism
FM	Ferromagnetic/ferromagnetism
FS	Fermi surface
HAFM	Helical antiferromagnetic/antiferromagnetism
hcp	Hexagonal close packed
HK	Hohenberg-Kohn (theorem)
HRE	Heavy rare earth
KKR	Korringa-Kohn-Rostoker (electronic structure method)
LDA	Local Density Approximation
LDA+U	Local Density Approximation + U (Hubbard parameter)
MCE	Magnetocaloric Effect
MST	Multiple Scattering Theory
MTA	Muffin-tin approximation
PM	Paramagnetic/paramagnetism
RKKY	Ruderman-Kittel-Kasuya-Yosida (interaction)
RSA	Rigid Spin Approximation

SDFT	Spin-Density Functional Theory
SIC	Self-interaction correction
SPO	Scattering path operator
TCE	Toroidocaloric Effect

Chapter 1
Introduction

Magnetism in a solid-state is a collective phenomenon arising from the interactions among an immense number of electrons, which carry an intrinsic degree of freedom known as spin, and the fixed nuclei. In magnetic materials the electrons act cooperatively to yield the formation of local magnetic moments, which are atom-scale spin polarisation of the electronic density and central magnetic entities at the sub-nanoscale. Magnetic moments can have different orientations and sizes at different atomic sites, and so ordered magnetic patterns called magnetic phases can stabilise following the complexity of the underlying electronic motions and their interactions. The range of magnetic orderings observed in nature is enormous and rich. For example, the simplest situation corresponds to ferromagnetism, in which all the magnetic moments are parallel aligned and a total non-vanishing magnetisation is produced. This order is in sharp contrast with the magnetism originated in antiferromagnets, where the net magnetic moment vanishes owing to the full compensation of the local moment orientations and sizes. Non-collinear helimagnetic spirals [1, 2] and Skyrmion structures [3–5] are some examples of the extremely complicated magnetic arrangement and patterns that can be established.

Such a large variety of magnetic phases has been extensively exploited in the design of technological devices and novel material functionality. Some recent examples of magnetic phenomena being used for technological applications are collinear and non-collinear antiferromagnetism for spintronics [6–8] and solid-state memories [9], in which both the charge and the magnetic moment—or spin—properties are simultaneously exploited. Refrigeration technology generated thanks to the presence of magnetism has been also the focus of study in the last decades. This technology has the potential to be disruptive in our society since it promises to be environmentally friendly and more efficient than current engines based on the compression and expansion of harmful gases [10, 11]. Both ferromagnetism [12] and antiferromagnetism [13, 14] are often considered to produce cooling effects, hugely expanding the available mechanisms to create new cooling devices. Frustrated antiferromagnetism, in which the geometrical arrangement of the atoms leads to a conflict of competing magnetic interactions between the magnetic moments, has also been exploited to obtain novel and complicated phases for their use to produce refrigeration [15]. In

© Springer Nature Switzerland AG 2020

E. Mendive Tapia, *Ab initio Theory of Magnetic Ordering*, Springer Theses,
https://doi.org/10.1007/978-3-030-37238-5_1

particular, results in this thesis show how magnetic frustration and magnetic phases with complex temperature-dependent properties can be used to enhance cooling efficiency.

As such condensed matter phases, the macroscopic properties of magnetic structures are the result of the complicated behaviour of their constituents, the magnetic moments. The overall magnetism can only be accurately described by understanding its quantum mechanical origins from an atomic point of view. This situation sets a very fundamental and complicated problem that still puzzles physicists today, even from what can be considered a very essential aspect: the description of magnetic moment formation and their interactions. It should not be regarded as surprising that this task is still not fully accomplished and that many fundamental questions remain without an answer. Firstly, magnetic moments, whose collective behaviour determines magnetic order, in turn emerge from the spins of many interacting electrons. This raises questions regarding the mutual influence between both cooperative mechanisms. Time and energy scale differences clearly play important roles in this mixing. Secondly, the spin of the electrons is a direct consequence of relativistic effects, hence fundamentally including the complexity of these into the intrinsic many-body quantum mechanical character of the problem. Furthermore, experience shows that relatively small temperature changes can have a profound impact on the properties of a magnetic material. For example, transitions between different kind of magnetic phases can be thermally triggered and heating up to high enough temperatures eventually destroys the overall magnetic order without suppressing the magnetic moments themselves. The effect of thermal fluctuations on magnetism and the underlying electronic glue, described by the sea of electrons binding together the nuclei, should therefore be also taken into account to attain a full description.

We present a theory to study the temperature-dependent magnetic properties of metallic materials from first principles, i.e. from a parameter-free formulation and relying on well established laws of nature only. Evidently, such a theory must be grounded in a sound quantum mechanical basis and ideally be fully extended to naturally include relativistic effects. Hence, we employ Density Functional Theory (DFT) [16, 17], a broadly used technique developed to solve the quantum mechanical equations of electrons in a solid within an effective single particle picture. It, therefore, allows to efficiently describe the electronic glue, or electronic structure, of a magnetic material. The choice of this method is not made just for its efficiency and versatility. It also permits a natural implementation of a suitable approach able to render magnetic disorder computationally tractable, which is necessary to describe different states of magnetic order and their dependence on temperature. The route to achieve this goes through the usage of Green's functions and Multiple Scattering Theory (MST), known as the Korringa–Kohn–Rostoker (KKR)-MST method [18, 19], in honour to its early developers. Such an approach distinguishes itself to other electronic structure techniques in that it uses the less intuitive Green's functions as main mathematical tools, instead of the more familiar wave functions often employed in other DFT-based calculations. This feature precisely allows to incorporate the formalism to describe magnetic moment fluctuations affecting and being influenced by the underlying electronic structure. A central tenet of the theory used in this work is to assume that the local moment orientations vary very slowly on the time-scale of the other electronic motions. This is the approach known as Disordered Local Moment (DLM) theory [20], which allows to treat magnetism and magnetic phase stability at finite temperatures very efficiently. DFT-DLM theory has been successfully implemented in the past and recent years to study, for example, the onset of magnetic order in strongly-correlated systems [21] and the heavy rare earth elements [2], metamagnetic phase transitions in metal alloys [14, 22], the magnetic interactions between rare-earth and transition-metals [12, 23], and temperature-dependent magnetic anisotropy [24–27].

An important part of this thesis is the use of DFT-DLM theory to investigate the effect of pairwise and higher order multi-spin interactions, and how they produce cooling mechanisms. A key aspect is to show how the dependence of the electronic structure of a magnetic material on magnetic order is linked to the presence of these multi-spin effects, i.e. terms that contain local moment interactions over groups of more than two magnetic atoms. As the magnetic ordering develops at each atomic site, the effect of multi-spin interactions might become important in systems with a complicated coupling between the electronic glue and the local moments, which can be expected to occur in metallic magnets, for instance. The work presented in this thesis shows that multi-spin effects are necessary to describe complex magnetic behaviour that depends on the state of magnetic order itself. The interest is to study how such a phenomenon can have an important impact on magnetic phase stability and transitions between different magnetic phases. This task is accomplished within the DLM approach by modelling transverse excitations of the magnetic moment orientations, since it naturally describes how the electronic structure spin-polarises and so adapts to the extent and type of magnetic order. A method developed to calculate pairwise and multi-spin interactions is presented, which uses a mean-field treatment to solve the statistical mechanics of the local moment orientations. A central outcome produced by this method is the free energy of a magnetic material, as well as magnetic moment order as function of temperature, magnetic field and interatomic distances. It also describes temperature and entropy changes between different magnetic states, thus being a natural tool for the calculation of cooling effects for their use in refrigeration technology exploiting magnetism [10, 28]. We also develop a methodology to construct magnetic phase diagrams containing second-(continuous) and first-order (discontinuous) magnetic phase transitions. Consequent tricritical points in which the magnetic phase transition changes from first- to second-order can be, therefore, located. Since the theory is able to track the order of the transitions, the examination of the magnetic phase diagrams can be also used to study previously unexplored phase-space regions with the most optimal conditions maximising thermodynamic quantities of interest. As it will be shown, both pair- and multi-spin interactions play an important role in this situation since they can determine a large part of the magnetic phase diagram's features.

The thesis is organised as follows. In Chap. 2 we give an overview of DFT and KKR-MST as well as of an effective medium theory of disorder naturally implementable within our techniques. The DLM approach and how it can be used together with DFT is presented in Chap. 3. A linear response theory for the fully disordered local moment regime, formally describing the fully disordered paramagnetic state in a rigorous way, is also provided in this chapter. This part of the work is of particular importance because it provides a scheme to identify the most stable and potential magnetic orderings in a magnetic material. Chapter 4, which mainly contains the original contribution in terms of theoretical developments, is devoted to show our methodology to obtain pairwise and multi-spin interactions and how these can be used to describe the stabilisation of magnetic phases against temperature, magnetic field and lattice positions. We also explain in this chapter how to construct magnetic phase diagrams. To demonstrate some of the particularities of the theory, we apply it to bcc iron and present some simple case studies for illustrative purposes. Chapters 5, 6, and 7 are dedicated to show new results, which also include a summary and conclusions at the end of each one of them. We first study in depth the incommensurate magnetism and temperature and magnetic field dependence of long-period magnetic phases, as well as ferromagnetism, in the heavy rare earth elements in Chap. 5. The effect of biaxial strain on the frustrated magnetism in the Mn-based antiperovskite system Mn_3GaN, as well as how the theory developed meets the challenge of accurately describing the magnetism of a long range of magnetic and lattice structures in the Mn_3A (A = Sn, Ga, Ge, Pt, Rh, Ir) materials class, are presented in Chaps. 6 and 7, respectively. These three chapters contain the original work published in references [29–31]. Finally, in Chap. 8 we give a summary together with an overview and discussion of future work.

References

1. Jensen J, Mackintosh AR (1991) Rare earth magnetism: structure and excitations. Clarendon, Oxford
2. Hughes ID, Däne M, Ernst A, Hergert W, Lüders M, Poulter J, Staunton JB, Svane A, Szotek Z, Temmerman WM (2007) Lanthanide contraction and magnetism in the heavy rare earth elements. Nature 446:650–653
3. Mühlbauer S, Binz B, Jonietz F, Pfleiderer C, Rosch A, Neubauer A, Georgii R, Böni P (2009) Skyrmion lattice in a chiral magnet. Science 323(5916):915–919
4. Nagaosa N, Tokura Y (2013) Topological properties and dynamics of magnetic skyrmions. Nat Nanotechnol 8:899
5. Kanazawa N, Seki S, Tokura Y (2017) Noncentrosymmetric magnets hosting magnetic skyrmions. Adv Mater 29(25):1603227
6. Jungwirth T, Marti X, Wadley P, Wunderlich J (2016) Antiferromagnetic spintronics. Nat Nanotechnol 11:231
7. Zhang Y, Sun Y, Yang H, Železný J, Parkin SPP, Felser C, Yan B (2017) Strong anisotropic anomalous hall effect and spin hall effect in the chiral antiferromagnetic compounds Mn_3X (X = Ge, Sn, Ga, Ir, Rh, and Pt). Phys Rev B 95:075128
8. Železný J, Zhang Y, Felser C, Yan B (2017) Spin-polarized current in noncollinear antiferromagnets. Phys Rev Lett 119:187204
9. Wadley P, Howells B, Železný J, Andrews C, Hills V, Campion RP, Novák V, Olejník K, Maccherozzi F, Dhesi SS, Martin SY, Wagner T, Wunderlich J, Freimuth F, Mokrousov Y, Kuneš J, Chauhan JS, Grzybowski MJ, Rushforth AW, Edmonds KW, Gallagher BL, Jungwirth T (2016) Electrical switching of an antiferromagnet. Science 351(6273):587–590

10. Tishin AM, Spichkin YI (2003) The magnetocaloric effect and its applications, vol 6. CRC Press, Boca Raton

11. Sandeman KG (2012) Magnetocaloric materials: the search for new systems. Scr Mater 67(6):566–571. Viewpoint Set No. 51: magnetic materials for energy

12. Petit L, Paudyal D, Mudryk Y, Gschneidner KA, Pecharsky VK, Lüders M, Szotek Z, Banerjee R, Staunton JB (2015) Complex magnetism of lanthanide intermetallics and the role of their valence electrons: Ab initio theory and experiment. Phys Rev Lett 115:207201

13. Nikitin SA, Myalikgulyev G, Annaorazov MP, Tyurin AL, Myndyev RW, Akopyan SA (1992) Giant elastocaloric effect in FeRh alloy. Phys Lett A 171(3):234–236

14. Staunton JB, Banerjee R, dos Santos DM, Deak A, Szunyogh L (2014) Fluctuating local moments, itinerant electrons, and the magnetocaloric effect: compositional hypersensitivity of FeRh. Phys Rev B 89:054427

15. Matsunami D, Fujita A, Takenaka K, Kano M (2015) Giant barocaloric effect enhanced by the frustration of the antiferromagnetic phase in Mn_3GaN. Nat Mater 14:73

16. Hohenberg P, Kohn W (1964) Inhomogeneous electron gas. Phys Rev 136:B864–B871

17. Kohn W, Sham LJ (1965) Self-consistent equations including exchange and correlation effects. Phys Rev 140:A1133–A1138

18. Korringa J (1947) On the calculation of the energy of a Bloch wave in a metal. Physica 13(6):392–400

19. Kohn W, Rostoker N (1954) Solution of the Schrödinger equation in periodic lattices with an application to metallic lithium. Phys Rev 94:1111–1120

20. Gyorffy BL, Pindor AJ, Staunton J, Stocks GM, Winter H (1985) A first-principles theory of ferromagnetic phase transitions in metals. J Phys F Met Phys 15(6):1337

21. Hughes ID, Däne M, Ernst A, Hergert W, Lüders M, Staunton JB, Szotek Z, Temmerman WM (2008) Onset of magnetic order in strongly-correlated systems from ab initio electronic structure calculations: application to transition metal oxides. New J Phys 10(6):063010

22. Staunton JB, dos Santos DM, Peace J, Gercsi Z, Sandeman KG (2013) Tuning the metamagnetism of an antiferromagnetic metal. Phys Rev B 87:060404

23. Patrick CE, Kumar S, Balakrishnan G, Edwards RS, Lees MR, Mendive-Tapia E, Petit L, Staunton JB (2017) Rare-earth/transition-metal magnetic interactions in pristine and (Ni, Fe)-doped YCo_5 and $GdCo_5$. Phys Rev Mater 1:024411

24. Staunton JB, Ostanin S, Razee SSA, Gyorffy BL, Szunyogh L, Ginatempo B, Bruno E (2004) Temperature dependent magnetic anisotropy in metallic magnets from an ab initio electronic structure theory: $L1_0$-ordered FePt. Phys Rev Lett 93:257204

25. Staunton JB, Szunyogh L, Buruzs A, Gyorffy BL, Ostanin S, Udvardi L (2006) Temperature dependence of magnetic anisotropy: an ab initio approach. Phys Rev B 74:144411

26. Matsumoto M, Banerjee R, Staunton JB (2014) Improvement of magnetic hardness at finite temperatures: Ab initio disordered local-moment approach for YCo_5. Phys Rev B 90:054421

27. Patrick CE, Kumar S, Balakrishnan G, Edwards RS, Lees MR, Petit L, Staunton JB (2018) Calculating the magnetic anisotropy of rare-earth-transition-metal ferrimagnets. Phys Rev Lett 120:097202

28. Planes A, Ll M, Saxena A (2014) Magnetism and structure in functional materials. Springer, Berlin

29. Mendive-Tapia E, Staunton JB (2017) Theory of magnetic ordering in the heavy rare earths: Ab initio electronic origin of pair- and four-spin interactions. Phys Rev Lett 118:197202

30. Zemen J, Mendive-Tapia E, Gercsi Z, Banerjee R, Staunton JB, Sandeman KG (2017) Frustrated magnetism and caloric effects in Mn-based antiperovskite nitrides: Ab initio theory. Phys Rev B 95:184438

31. Mendive-Tapia E, Staunton JB (2019) Ab initio theory of the gibbs free energy and a hierarchy of local moment correlation functions in itinerant electron systems: the magnetism of the mn_3a materials class. Phys Rev B 99:144424

Chapter 2
Ab initio Theory of Electronic Structure

Describing the very complicated process of solid formation initially demands to fully solve the quantum many-body problem of coupled electrons and nuclei. This is a formidable task that must be simplified in order to be tractable. One can adopt the approach of designing a simpler Hamiltonian aimed to model a small number of degrees of freedom of interest, and let the rest to be captured by empirical parameters. This path fundamentally relies on experimental experience and adjusting the Hamiltonian to the specifications of each system is usually necessary. Alternatively, one refers to ab initio modelling, or equivalently a model from 'first principles', when the theory used does not rely on empirical parameters and only free-parameter approximations are incorporated. Such a theory is evidently very desirable and many approaches have been produced and developed in the last decades. Along these lines, the development of the well established Density Functional Theory (DFT) technique, which led to the Nobel prize in chemistry in 1998, represented a breakthrough since it provided the basis to study solids from purely theoretical inputs. DFT has been exploited to study very diverse phenomena and has substantially advanced the understanding in many fields, such as physics, chemistry, and engineering. Due to its reliability and relative simplicity, it has become one of the most employed ab initio approaches and its study is an established field which is still continuously evolving yet today.

This chapter is devoted to introduce DFT, which will be used as the ab initio technique supporting our calculations to describe the electronic structure. Firstly, in Sect. 2.1 we derive the central Kohn–Sham equations behind DFT, present the approximation made for the calculation of the exchange and correlation functional, and show the formalism of DFT at non-zero temperatures. The formal solution of the Kohn–Sham equations, based on the multiple scattering approach, is then presented and developed in Sect. 2.2. Finally, in Sect. 2.3 we introduce a theory to describe disorder by constructing an effective medium.

© Springer Nature Switzerland AG 2020

E. Mendive Tapia, *Ab initio Theory of Magnetic Ordering*, Springer Theses,
https://doi.org/10.1007/978-3-030-37238-5_2

2.1 Density Functional Theory

DFT is aimed to make the very complicated problem of many interacting electrons surrounded by many nuclei tractable. The problem can be initially simplified by considering that in the solid state the nuclei typically remain at fixed positions, or barely move from them, compared with the fast travelling electrons. This is due to the fact that their mass is about three orders of magnitude larger than the mass of the electrons and, consequently, the electronic velocities must be much smaller. From this, one can neglect the kinetic energy of the nuclei and consider the effect of their presence as an external potential acting on the electrons. This is the so-called Born–Oppenheimer approximation [1]. In this situation the electrons feel the effect of a stationary effective field created by the slowly moving underlying nuclei arrangement and instantaneously adjust their motions to its evolution. The non-relativistic Hamiltonian operator describing this picture can be written as [2]

$$\hat{\mathcal{H}} = \hat{K} + \hat{W} + \sum_n V_{\text{ext}}(\mathbf{r}_n),$$

$$\hat{K} = -\frac{\hbar^2}{2m_e} \sum_n \nabla^2_{\mathbf{r}_n} \qquad \hat{W} = \frac{1}{2} \sum_n \sum_{m \neq n} \frac{e^2}{4\pi\varepsilon_0 |\mathbf{r}_n - \mathbf{r}_m|}, \tag{2.1}$$

where ε_0 is the permittivity of vacuum and e, m_e, and $\mathbf{r}_n$ are the charge, mass and position of the nth electron, respectively. The term labelled as $\hat{K}$ is the kinetic energy, and the external potential V_{ext} comprises the effect of the fixed nuclei on the electrons. The second term $\hat{W}$ describes the Coulomb repulsion between electron pairs and, in practice, its presence makes the Hamiltonian in Eq. (2.1) too complicated to be solved as it stands, since it represents the source of the many-electron interaction. The strategy behind DFT to overcome this problem consists in converting Eq. (2.1) into an independent electron Hamiltonian. This idea is implemented by firstly introducing the central concept of DFT established by Hohenberg and Kohn: the total energy of the system is a functional of the electron density $n(\mathbf{r})$ only,

$$E = E[n(\mathbf{r})]. \tag{2.2}$$

This is the most important statement of the Hohenberg–Kohn (HK) theorem and ensures that the total energy is minimised by the ground state electron density $n_0(\mathbf{r})$ [3]. This observation simplifies the problem remarkably since it shows that there is only need to find $n(\mathbf{r})$, instead of the immense many-body wave function $\Psi(\mathbf{r}_1, \mathbf{r}_2, \dots)$, in order to calculate the total energy at the ground state. Returning now to Eq. (2.1), the total energy functional is expressed as

$$E[n] = \langle \Psi[n] | \hat{K} + \hat{W} | \Psi[n] \rangle + \int d\mathbf{r}\, n(\mathbf{r}) V_{\text{ext}}(\mathbf{r}) = F[n] + \int d\mathbf{r}\, n(\mathbf{r}) V_{\text{ext}}(\mathbf{r}),$$

$$\tag{2.3}$$

where $|\Psi\rangle$ is the many-body ground state wave function and $F[n] = K[n] + W[n]$ is a universal functional which does not depend on the particularities of the crystal structure. The Kohn–Sham approach now follows by splitting $F[n]$ into kinetic K_0, and Hartree E_H, energy terms of an effective independent electron system reproducing exactly the density of the original one,

$$F[n] = K_0[n] + E_H[n] + E_{xc}[n]. \tag{2.4}$$

Here

$$K_0[n] = -\frac{\hbar^2}{2m} \sum_i \int \mathrm{d}\mathbf{r}\,\psi_i^*(\mathbf{r})\nabla^2\psi_i(\mathbf{r})$$

$$E_H[n] = \frac{1}{2}\int\int \mathrm{d}\mathbf{r}\mathrm{d}\mathbf{r}' n(\mathbf{r})\frac{e^2}{4\pi\varepsilon_0|\mathbf{r} - \mathbf{r}'|}n(\mathbf{r}') = \frac{1}{2}\int \mathrm{d}\mathbf{r} n(\mathbf{r})V_H(\mathbf{r}) \tag{2.5}$$

defines the Hartree potential $V_H(\mathbf{r})$ and introduces the Kohn–Sham single-particle wave functions $\{\psi_i(\mathbf{r})\}$, yet unknown, of the independent electron Hamiltonian. The extra energy term $E_{xc}[n]$ in Eq. (2.4), named exchange and correlation energy, is also unknown and accounts for the difference between the non-interacting effective medium and the many-body real system,

$$E_{xc}[n] = K[n] + W[n] - K_0[n] - E_H[n]. \tag{2.6}$$

In practice, this term has to be approximated and ideally it is expected to be relatively small. Evidently, the accuracy of DFT calculations largely rely on how this term is treated.

The electron density is computed from the single electron wave functions as

$$n(\mathbf{r}) = \sum_i |\psi_i(\mathbf{r})|^2, \tag{2.7}$$

where the summation is over occupied states, and spin degrees of freedom if relevant to the problem. An important premise of the HK theorem is that $n_0(\mathbf{r})$ determines uniquely the external potential of the nuclei $V_{ext}(\mathbf{r})$ [3, 4]. In fact this one-to-one mapping does not depend on the explicit form of the electron-electron interaction and, in consequence, it guarantees the existence of a non-interacting system which has the same ground state electron density as the real system. In order to obtain the wave functions $\{\psi_n(\mathbf{r})\}$ it is convenient to introduce the Lagrange functional

$$L[n] = E[n] - \sum_{ij} \lambda_{ij} \left[\langle\psi_i(\mathbf{r})|\psi_j(\mathbf{r})\rangle - \delta_{ij} \right], \tag{2.8}$$

where λ_{ij} are the corresponding Lagrange multipliers ensuring the orthogonality of the wave functions. The variational principle is then used to minimise the total energy and simultaneously constrain the calculation to the correct number of electrons in the system, N. From Eq. (2.8) one then obtains

$$\frac{\delta L[n]}{\delta n} = 0 \Rightarrow \frac{\delta E[n]}{\delta \psi_i^*(\mathbf{r})} - \sum_j \lambda_{ij} \psi_j(\mathbf{r}) = 0. \tag{2.9}$$

After some simple algebra and using Eqs. (2.3), (2.4), and (2.5) one can write

$$\left[-\frac{\hbar^2 \nabla^2}{2m_e} + V_{\text{ext}}(\mathbf{r}) + V_{\text{H}}(\mathbf{r}) + \frac{\delta E_{\text{xc}}[n]}{\delta n} \right] \psi_i(\mathbf{r}) = \sum_j \lambda_{ij} \psi_j(\mathbf{r}), \tag{2.10}$$

which can be rewritten in terms of rotated single-particle wave functions $\{\phi_i(\mathbf{r})\}$, from diagonalizing λ_{ij}, and the exchange and correlation potential is defined as

$$V_{\text{xc}}(\mathbf{r}) = \left. \frac{\delta E_{\text{xc}}[n]}{\delta n} \right|_{n(\mathbf{r})}, \tag{2.11}$$

that is,

$$\left[-\frac{\hbar^2 \nabla^2}{2m_e} + V_{\text{ext}}(\mathbf{r}) + V_{\text{H}}(\mathbf{r}) + V_{\text{xc}}(\mathbf{r}) \right] \phi_i(\mathbf{r}) = \varepsilon_i \phi_i(\mathbf{r}), \tag{2.12}$$

where $\{\varepsilon_i\}$ are the single-particle energies. This set of equations are known as the Kohn–Sham equations [4] and are the basis of DFT and the cornerstone of the most important modern computational electronic structure theories for modelling materials ab initio. Note that the effective potential finally derived in Eq. (2.12)

$$V_{\text{eff}}(\mathbf{r}) = V_{\text{ext}}(\mathbf{r}) + V_{\text{H}}(\mathbf{r}) + V_{\text{xc}}(\mathbf{r}) \tag{2.13}$$

depends on $n(\mathbf{r})$ so that the Kohn–Sham equations can be solved self-consistently together with Eq. (2.7). An initial guess of the electron density is taken to calculate $V_{\text{eff}}(\mathbf{r})$ and the single-particle wave functions are obtained by solving the Kohn–Sham equations. At this point these solutions are used to recalculate the electron density. If one obtains a different value then a new guess is constructed, using the knowledge of the previous potentials, and the cycle is repeated until self-consistency is reached. The total energy can be calculated too using the single-particle energies $\{\varepsilon_n\}$ as

$$E[n] = \sum_i f_i \varepsilon_i - E_{\text{H}}[n] + E_{\text{xc}}[n] - \int d\mathbf{r} n(\mathbf{r}) V_{\text{xc}}(\mathbf{r}), \tag{2.14}$$

where $f_i = 1$ ($f_i = 0$) applies if the state is occupied (non-occupied) and the extra terms are added to subtract double-counting terms in the Hartree and the exchange-correlation contributions.

2.1.1 The Local Density Approximation

As explained in the previous section, the exchange-correlation potential is designed to capture all the complicated many-body effects initially ignored within the independent electron approximation. Until today the corresponding electron density functional $E_{xc}[n]$ is still unknown and continuous effort to construct accurate exchange and correlation potentials constitutes an active research field itself. Here we employ the most common numerical scheme to approximate $E_{xc}[n]$, the Local Density Approximation (LDA) [5, 6]. The strategy is to calculate the exchange and correlation energies of a simpler system, the interacting homogeneous electron gas, and map them locally to the real medium. If the electron density is slowly varying, one can assume that $E_{xc}[n]$ in each infinitesimal volume element at a given position is the same as the one calculated for the homogeneous electron gas with the same value of the charge density [2],

$$\mathrm{d}E_{xc} = \frac{E_{xc}^{hom}[n(\mathbf{r})]}{V}\mathrm{d}\mathbf{r} \Rightarrow E_{xc}[n] = \frac{1}{V}\int \mathrm{d}\mathbf{r}\,E_{xc}^{hom}[n(\mathbf{r})], \tag{2.15}$$

where V is the volume and $E_{xc}[n]$ is then obtained by adding up all the individual contributions from each volume element. For the homogeneous electron gas the exchange functional can be derived exactly [7], while the correlation functional is extracted by solving numerically the many-particle Hamiltonian and parametrising the resulting data [5]. The total exchange-correlation functional is then obtained by summing up both contributions. In this thesis we implement the LDA and use the parametrisation proposed by Perdew and Wang [8].

The success of LDA is usually attributed to the fact that it satisfies some sum rules derived from the pair density and connected to the exchange-correlation hole [9]. The LDA is also partly motivated by the fact that $E_{xc}[n]$ should be a universal functional. It is expected, therefore, that both the exchange and correlation effects are well described by the Coulomb repulsion and the quantum statistical nature included in the homogeneous electron gas.

2.1.2 Density Functional Theory Extended to Finite Temperatures

Hohenberg and Kohn laid the basis of ab initio ground state calculations for the electronic structure of materials. To accomplish this they demonstrated the existence of the universal functional $F[n]$, independent of the external potential $V_{ext}(\mathbf{r})$, such that $E[n]$ in Eq. (2.3) is minimised for the ground state density $n_0(\mathbf{r})$ associated to $V_{ext}(\mathbf{r})$ [3]. However, finite temperature effects on the population of excited electronic states was not considered. The extension of DFT to finite temperatures was realised by Mermin, soon after the publication of Hohenberg and Kohn, by following an analogous reasoning in the context of the grand canonical ensemble [10]. The starting point is to construct the grand potential functional

$$\Omega[\hat{\rho}] = \text{Tr}\hat{\rho}\left(\hat{\mathcal{H}} - \nu\hat{N} + \frac{1}{\beta}\log\hat{\rho}\right), \tag{2.16}$$

where ν is the chemical potential, $\hat{N}$ is the particle number operator, and $\hat{\mathcal{H}}$ is the many-body Hamiltonian as expressed in Eq. (2.1). Equation (2.16) naturally introduces the probability density operator $\hat{\rho}$ such that $\text{Tr}\,[\ldots]$ performs the trace operation and the expected value of an operator $\hat{A}$ is calculated as $\langle\hat{A}\rangle = \text{Tr}\,\hat{\rho}\hat{A}$. The grand potential

$$\Omega = -\frac{1}{\beta}\log\left(\text{Tr}\exp\left[-\beta(\hat{\mathcal{H}} - \nu\hat{N})\right]\right) \tag{2.17}$$

is given by Eq. (2.16) when $\rho = \rho_0$ is the appropriate equilibrium density matrix

$$\rho_0 = \frac{\exp\left[-\beta(\hat{\mathcal{H}} - \nu\hat{N})\right]}{\text{Tr}\exp\left[-\beta(\hat{\mathcal{H}} - \nu\hat{N})\right]}. \tag{2.18}$$

The key point is that it is possible to show that in fact the minimum of $\Omega[\rho]$ is attained by ρ_0 and, therefore, $V_{\text{ext}}(\mathbf{r})$ determines ρ_0 [10]. Underlying the work of Mermin is to show that the functional variable can be transferred from ρ to $n(\mathbf{r})$ such that a universal functional $F_T[n]$ independent of $V_{\text{ext}}(\mathbf{r})$ exists and that

$$\Omega = \int \text{d}\mathbf{r}n_0(\mathbf{r})V_{\text{ext}}(\mathbf{r}) + F_T[n_0] \tag{2.19}$$

is the grand potential too, i.e., it is minimised by the equilibrium density $n_0(\mathbf{r})$ in the presence of $V_{\text{ext}}(\mathbf{r})$. Similarly to zero temperature DFT, it is straightforward to show that $V_{\text{ext}}(\mathbf{r})$ is unequivocally determined by $n_0(\mathbf{r})$ by *reductio ad absurdum* demonstration [10]. Since $V_{\text{ext}}(\mathbf{r})$ in turn determines ρ_0, it can be deduced that ρ_0 is a functional of $n_0(\mathbf{r})$. From Eqs. (2.16) and (2.19), and recalling the Hamiltonian in Eq. (2.1), it follows that

$$F_T[n_0] = \text{Tr}\left[\rho_0[n_0]\left(\hat{K} + \hat{W} + \frac{1}{\beta}\log\rho_0[n_0]\right)\right]. \tag{2.20}$$

Equation (2.19) is, therefore, the grand potential, showing that the approach presented by Hohenberg and Kohn can be implemented at non-zero temperatures.

In practice, the total energy expression given in Eq. (2.14) can be extended to non-zero temperatures by using the equilibrium electron density and the appropriate Fermi–Dirac occupation function $f_n = \left(1 + e^{\beta(\varepsilon_n - \nu)}\right)^{-1}$ to calculate

$$\Omega[n] = \Omega_0[n] - E_{\mathrm{H}}[n] + \Omega_{\mathrm{xc}}[n] - \int \mathrm{d}\mathbf{r} n(\mathbf{r}) V_{\mathrm{xc}}(\mathbf{r}), \tag{2.21}$$

where now the Hohenberg–Kohn energy functional becomes a Helmholtz free energy such that $\Omega_0[n] \to \sum_n f_n \varepsilon_n - T S_0 - \nu N$, which can be described by the non-interacting fermions entropy

$$S_0 = -k_{\mathrm{B}} \sum_n \Big[f_n \log f_n + (1 - f_n) \log(1 - f_n) \Big]. \tag{2.22}$$

In Chap. 3 we will show how this extension to finite temperature can be used to describe magnetic thermal fluctuations within a disordered local moment picture.

2.1.3 Magnetism and Relativistic Effects in Density Functional Theory

The DFT approach presented in the previous sections does not depend on the spin since its presence has been ignored so far for simplicity. The concept of spin is a direct consequence of the extension of the Schrödinger equation to be invariant with respect to Lorentz transformations. In other words, it arises naturally in the framework of the Dirac equation. Similarly, the generalisation of DFT to take in magnetism goes through the consideration of special relativity effects. Pertinent to the spin inclusion, therefore, is to examine the Dirac equation for an electron moving under the effect of an external magnetic field specified by a vector potential, i.e. $\mathbf{B} = \nabla \times \mathbf{A}(\mathbf{r})$,

$$\hat{\mathcal{H}}_D \Psi_i(\mathbf{r}) = \Big[c\boldsymbol{\alpha} \cdot \big(-i\hbar\nabla + e\mathbf{A}(\mathbf{r}) \big) + \beta_I m_e c^2 \Big] \Psi_i(\mathbf{r}) = \varepsilon_i \Psi_i(\mathbf{r}). \tag{2.23}$$

Here c is the velocity of light and ε_i is the energy solution. β_I is a four by four matrix defined as

$$\beta_I = \begin{pmatrix} I_2 & (0) \\ (0) & -I_2 \end{pmatrix}, \quad \text{where} \quad I_2 = \begin{pmatrix} 1 & 0 \\ 0 & 1 \end{pmatrix}, \tag{2.24}$$

(0) being a matrix of zeros, and $\boldsymbol{\alpha}$ is a vector whose components are defined as

$$\boldsymbol{\alpha} = (\alpha_x, \alpha_y, \alpha_z) \quad \text{with} \quad \alpha_i = \begin{pmatrix} (0) & \sigma_i \\ \sigma_i & (0) \end{pmatrix}, \tag{2.25}$$

where $\{\sigma_i\}$ are the well known Pauli matrices

$$\sigma_x = \begin{pmatrix} 0 & 1 \\ 1 & 0 \end{pmatrix}, \quad \sigma_y = \begin{pmatrix} 0 & -i \\ i & 0 \end{pmatrix}, \quad \sigma_z = \begin{pmatrix} 1 & 0 \\ 0 & -1 \end{pmatrix}. \tag{2.26}$$

The solution of Eq. (2.23), $\boldsymbol{\Psi}_i(\mathbf{r})$, is the famous Dirac spinor, which is structured by four different components. Hence, it is natural to expect that in relativistic DFT the central role played by the electron density is replaced by a four-component function. Indeed, it has been shown that the total energy of the Dirac equation at the ground state is a unique functional of the relativistic four-component current $J_\mu(\mathbf{r})$ [11]. This is not surprising as $J_\mu(\mathbf{r})$ conceals the electron density $n(\mathbf{r})$ itself as well as the spin density. It includes the electron current density too, although it is usually ignored if diamagnetism and electric polarisation effects are not the object of interest. If the starting point of the calculation is the Dirac equation and spin-polarisation effects are included we refer to Spin-Density Functional Theory (SDFT). Importantly, the extension to finite temperatures of SDFT can be carried out too by the same argument presented by Mermin [10] since the mapping of the probability density to $J_\mu(\mathbf{r})$ can be equivalently argued. As it is common practice, in this thesis we focus our attention to the electron and spin densities only.

From energy comparison arguments and neglecting diamagnetism and orbital paramagnetism, Eq. (2.23) can be safely approximated to the following two- component equation [2]

$$\left[-\frac{\hbar^2 \nabla^2}{2m_e} + \frac{2\mu_B}{\hbar} \mathbf{S} \cdot \mathbf{B} \right] \boldsymbol{\Phi}_i(\mathbf{r}) = \varepsilon_i \boldsymbol{\Phi}_i(\mathbf{r}). \tag{2.27}$$

This equation is referred to as the Pauli equation and shows how the spin, described by the spin operator

$$\mathbf{S} = \frac{\hbar}{2}\boldsymbol{\sigma}, \quad \text{where} \quad \boldsymbol{\sigma} = (\sigma_x, \sigma_y, \sigma_z), \tag{2.28}$$

appears in the Dirac equation naturally. The Dirac spinor is now formed by independent two-spinor functions

$$\left\{ \boldsymbol{\Phi}_i(\mathbf{r}) = \phi_i(\mathbf{r}; 1)\, |\uparrow\rangle + \phi_i(\mathbf{r}; 2)\, |\downarrow\rangle \right\}, \quad \text{with} \quad |\uparrow\rangle = \begin{pmatrix} 1 \\ 0 \end{pmatrix} \quad \text{and} \quad |\downarrow\rangle = \begin{pmatrix} 0 \\ 1 \end{pmatrix}. \tag{2.29}$$

As usual, the electron density is constructed as the sum of the independent densities over occupied states

$$n(\mathbf{r}) = \sum_i n_i(\mathbf{r}) = \sum_i \boldsymbol{\Phi}_i^\dagger(\mathbf{r})\boldsymbol{\Phi}_i(\mathbf{r}) = \sum_i \left(|\phi_i(\mathbf{r}; 1)|^2 + |\phi_i(\mathbf{r}; 2)|^2 \right), \tag{2.30}$$

such as the total number of electrons satisfies $N_e = \int d\mathbf{r}\, n(\mathbf{r})$. The total spin density is computed from the application of the spin operator,

$$s(\mathbf{r}) = \sum_i s_i(\mathbf{r}) = \sum_i \boldsymbol{\Phi}_i^\dagger(\mathbf{r}) \mathbf{S} \boldsymbol{\Phi}_i(\mathbf{r}), \tag{2.31}$$

which suggests the following definition of the associated total magnetic moment density

$$\boldsymbol{\mu}(\mathbf{r}) = -\frac{2\mu_B}{\hbar} \sum_i \mathbf{s}_i(\mathbf{r}) = -\frac{2\mu_B}{\hbar} \mathbf{s}(\mathbf{r}). \tag{2.32}$$

From the definition given in Eq. (2.28), together with Eq. (2.29), we can write after some simple algebra the dependence of the spin density on the Dirac solution,

$$s_x(\mathbf{r}) = \hbar \operatorname{Re} \sum_i \left[\phi_i^*(\mathbf{r}; 1)\phi_i(\mathbf{r}; 2) \right], \tag{2.33}$$

$$s_y(\mathbf{r}) = \hbar \operatorname{Im} \sum_i \left[\phi_i^*(\mathbf{r}; 1)\phi_i(\mathbf{r}; 2) \right], \tag{2.34}$$

$$s_z(\mathbf{r}) = \frac{\hbar}{2} \sum_i \left[|\phi_i(\mathbf{r}; 1)|^2 - |\phi_i(\mathbf{r}; 2)|^2 \right], \tag{2.35}$$

which satisfies $|\mathbf{s}(\mathbf{r})| = \frac{\hbar}{2} n(\mathbf{r})$. An important observation regarding SDFT follows from this; we associate a three-dimensional spin polarisation at every point in space where the electron density is non-zero, which can be computed directly from the wave function components. Note that the magnetisation orientation is not necessarily collinear with respect to other spatial positions and can vary spatially in a non-trivial manner. Within this framework the total local magnetic moment per unit cell of volume V_u is obtained by performing the integral

$$\boldsymbol{\mu}_{V_u} = \int_{V_u} d\mathbf{r}\, \boldsymbol{\mu}(\mathbf{r}). \tag{2.36}$$

We return now to the density functional strategy of DFT and proceed to apply it to the Dirac equation. Clearly, once relativistic effects are included the first step is to consider the total energy as a functional of both the electron density and the magnetic moment density, i.e. $E[n(\mathbf{r})] \rightarrow E[n(\mathbf{r}), \boldsymbol{\mu}(\mathbf{r})]$. Adapting the Dirac equation to the approximations cementing DFT, namely the Born–Oppenheimer, mean-field (in the Hartree potential), and single-particle approximations, the total energy can be expressed in terms of the kinetic, exchange-correlation, and Hartree energies already introduced for non-relativistic DFT, as well as an extra contribution accounting for the external magnetic field

$$E[n, \boldsymbol{\mu}] = K_0[n, \boldsymbol{\mu}] + E_H[n, \boldsymbol{\mu}] + E_{xc}[n, \boldsymbol{\mu}] + \int d\mathbf{r}\left(n(\mathbf{r}) V_{ext}(\mathbf{r}) - \mathbf{B} \cdot \boldsymbol{\mu}(\mathbf{r}) \right). \tag{2.37}$$

The SDFT Kohn–Sham equations are then derived by following an analogous minimisation procedure with respect to $n(\mathbf{r})$ and $\boldsymbol{\mu}(\mathbf{r})$, as described in Sect. 2.1. For this purpose we define the density matrix

$$n_{\alpha\beta}(\mathbf{r}) = \sum_i \phi_i^*(\mathbf{r}; \alpha)\phi_i(\mathbf{r}; \beta), \tag{2.38}$$

which compactly contains both densities,

$$n(\mathbf{r}) = \sum_\alpha n_{\alpha\alpha}(\mathbf{r}) \quad \text{and} \quad \boldsymbol{\mu}(\mathbf{r}) = -\mu_{\mathrm{B}} \sum_{\alpha\beta} n_{\alpha\beta}(\mathbf{r})\boldsymbol{\sigma}_{\alpha\beta}. \tag{2.39}$$

The minimum principle follows, therefore, as

$$\frac{\delta E[n_{\alpha\beta}]}{\delta n_{\alpha\beta}} = 0, \tag{2.40}$$

which, together with the use of the Lagrange parameter associated with the particle number conservation, leads to the SDFT version of the Kohn–Sham equations

$$\left[-\frac{\hbar^2}{2m_e}\nabla^2 + V_{\text{ext}}(\mathbf{r}) + V_{\mathrm{H}}(\mathbf{r}) + V_{\text{xc}}(\mathbf{r}) - \mu_{\mathrm{B}}\boldsymbol{\sigma}\cdot(\mathbf{B}(\mathbf{r}) + \mathbf{B}_{\text{xc}}(\mathbf{r})) \right] \Phi_i(\mathbf{r}) = \varepsilon_i \Phi_i(\mathbf{r}), \tag{2.41}$$

where

$$\frac{\delta E_{\text{xc}}[n_{\alpha\beta}]}{\delta n_{\alpha\beta}}\bigg|_{n(\mathbf{r}),\mu(\mathbf{r})} = V_{\text{xc}}(\mathbf{r})I_2 + \mu_{\mathrm{B}}\boldsymbol{\sigma}\cdot\mathbf{B}_{\text{xc}}(\mathbf{r}). \tag{2.42}$$

This equation shows that the very complicated many-electron physics, captured by the exchange-correlation energy, might generate an effective magnetic field $\mathbf{B}_{\text{xc}}(\mathbf{r})$ at every point in space. Evidently, this term comprises the effect of Pauli exclusion principle and the Coulomb interaction fundamentally behind magnetism and in principle captured by the LDA. It is this term, therefore, the one that is responsible of magnetic ordering if the appropriate conditions favouring the generation of spin polarisation are present.

The total energy can be calculated from Eq. (2.41) in analogy to the non-relativistic computation shown in Eq. (2.14),

$$E[n, \mu] = \sum_i f_i\varepsilon_i - E_{\mathrm{H}}[n] + E_{\text{xc}}[n, \mu] - \int d\mathbf{r}\, n(\mathbf{r})V_{\text{xc}}(\mathbf{r}) - \int d\mathbf{r}\,\boldsymbol{\mu}(\mathbf{r})\cdot(\mathbf{B}(\mathbf{r}) + \mathbf{B}_{\text{xc}}(\mathbf{r})) \tag{2.43}$$

Note that the main difference is that a term accounting for the contribution of the total magnetic field is added.

2.2 Multiple Scattering Theory

The Kohn–Sham equations lay the basis for the solution of the electronic structure of periodic solids, as presented in the previous section. Owing to the functional character of the theory, self-consistent calculations can be used to solve these equations and the electron and magnetic moment densities, together with the potentials, can be obtained in an iterative cycle procedure. Numerous methods can be employed to find the solution of Eq. (2.12), or of its relativistic version in Eq. (2.41), although their form naturally suggests to choose eigenvalue-based strategies plus some advantageous description of the wave function. In this thesis we follow an alternative route that might be conceptually more difficult but which offers some benefits in compensation, the so-called Multiple Scattering Theory (MST). The complication is due to the fact that this theory is formulated in terms of Green's functions instead of the more familiar wave solutions. Green's function based technology applied to periodic solids was suggested firstly by Korringa [12] and later by Kohn and Rostoker [13], baptising the multiple scattering approach as KKR method. However, KKR-MST methodology did not become popular until a couple of years later when its formulation in multiple scattering terms was made [14] and the scattering path operator was introduced [15]. As it will be apparent later in this section, the advantage is that single-site scattering events can be solved independently of the geometry of the system so that the problem can be conveniently separated into two parts.

In practice, The KKR-MST formalism and mechanism is exactly the same for both relativistic and non-relativistic approaches. The main difference lies in that the size of the central quantities, such as the Green's functions and scattering matrices, become higher within the relativistic formulation due to the additional spin-dimensionality. The shapes of the spherical expansions, solving the scattering problems, are consequently different too, but the central equations and relations defining the theory remain the same. For this reason, we present in the following pages the non-relativistic derivation for illustrative purposes and to avoid the mathematical complication added in the relativistic picture. This is appropriate since the relevant non-relativistic results can then be healed to incorporate relativistic effects by suitably tracing with respect to the spin degrees. The interested reader can find detailed derivations of the relativistic formulation in many references, such as [16–19].

2.2.1 Green's Functions and the Single-Center Scattering Problem

In this section the Green's function formulation is presented and used to solve the problem of one particle being scattered by a single localised potential. The results obtained will become the basis to expand the theory to deal with a collection of scattering potentials in Sect. 2.2.2.

We proceed by recalling the effective potential derived in Eq. (2.13). Our starting point is to approximate it as a sum of non-overlapping spherical potentials centred at fixed positions $\{\mathbf{R}_n\}$,

$$V_{\text{eff}}(\mathbf{r}) \approx \sum_n V_n(r_n), \quad \text{with } \mathbf{r}_n = \mathbf{r} - \mathbf{R}_n, \tag{2.44}$$

where $r_n = |\mathbf{r}_n|$, and $V_n(r_n) = 0$ for $r_n > r_{\mathrm{MT},n}$, $r_{\mathrm{MT},n}$ being named the muffin-tin radius of the spherical region n. Note that $r_{\mathrm{MT},n}$ can be different at each site. This form to split $V_{\mathrm{eff}}(\mathbf{r})$ is known as the muffin-tin approximation (MTA) and the region of zero potential outside the spheres is referred to as the interstitial region. Many authors use, additionally, other manners to approximate $V_{\mathrm{eff}}(\mathbf{r})$. For example, in the Atomic Sphere Approximation (ASA) the interstitial region is removed by increasing the volume of the spheres and allowing the consequent superposition of the potentials. However, the MST presented here strictly demands that the potentials at each site do not overlap. Although one might desire to remove the interstitial region as much as possible in order to minimise the effect of the approximation, in principle the ASA should be treated carefully if applied.

In this picture the Kohn–Sham Hamiltonian $\hat{\mathcal{H}}_{\mathrm{KS}} = -\hbar^2\nabla^2/2m_e + V_{\mathrm{eff}}(\mathbf{r})$ is meant to describe a particle travelling across an ensemble of scattering centres that are associated to the potentials created at each site of the lattice system. The solution of the single-centre scattering event at site n, which we denote with $\varphi_n(\mathbf{r})$, is described by the single-site Hamiltonian equation

$$\hat{\mathcal{H}}_n(\mathbf{r})\varphi_n(\mathbf{r}) = \left(-\frac{\hbar^2\nabla^2}{2m_e} + V_n(r_n)\right)\varphi_n(\mathbf{r}) = E\varphi_n(\mathbf{r}). \qquad (2.45)$$

From this equation the Green's function associated to $\hat{\mathcal{H}}_n(\mathbf{r})$ is formally defined as

$$G_n(E) = \lim_{\epsilon \to \pm 0}\left(E - \hat{\mathcal{H}}_n + i\epsilon\right)^{-1}, \qquad (2.46)$$

whose real space representation version satisfies

$$(E - \hat{\mathcal{H}}_n(\mathbf{r}))G_n(\mathbf{r}, \mathbf{r}', E) = \delta(\mathbf{r} - \mathbf{r}'). \qquad (2.47)$$

Similarly, the free-particle Green's function $G_0(\mathbf{r}, \mathbf{r}', E)$ is defined by replacing $\hat{\mathcal{H}}_n$ by the free-particle Hamiltonian $\hat{\mathcal{H}}_0(\mathbf{r}) = -\hbar^2\nabla^2/2m_e$ in Eq. (2.45). The advantage of using Green's functions is that the free-particle trajectory can be connected to the scattered solution. For example, the well known Dyson equation expresses $G_n(\mathbf{r}, \mathbf{r}', E)$ in terms of $G_0(\mathbf{r}, \mathbf{r}', E)$ as [16],

$$G_n(\mathbf{r}, \mathbf{r}', E) = G_0(\mathbf{r}, \mathbf{r}', E) + \int d\mathbf{r}'' G_0(\mathbf{r}, \mathbf{r}'', E)V_n(\mathbf{r}'' - \mathbf{R}_n)G_n(\mathbf{r}'', \mathbf{r}, E)$$

$$= G_0(\mathbf{r}, \mathbf{r}', E) + \int\int d\mathbf{r}'' d\mathbf{r}_2 G_0(\mathbf{r}, \mathbf{r}'', E)t_n(\mathbf{r}'', \mathbf{r}_2', E)G_0(\mathbf{r}_2, \mathbf{r}', E). \qquad (2.48)$$

The previous expression defines the real space representation of the very useful t-matrix function, $t_n(\mathbf{r}, \mathbf{r}', E)$. It is this quantity the one that we are interested to calculate since it fully describes the scattering effect from a single potential. Importantly, it satisfies a similar version of the Dyson equation itself[1]

---

[1] This can be easily shown from Eq. (2.48).

$$t_n(\mathbf{r}, \mathbf{r}', E) = V_n(r_n)\delta(\mathbf{r} - \mathbf{r}') + \int d\mathbf{r}_1 V_n(r_n)G_0(\mathbf{r}, \mathbf{r}_1, E)t_n(\mathbf{r}_1, \mathbf{r}', E). \quad (2.49)$$

In addition, the scattered wave function is connected to the free electron solution $\phi_n(\mathbf{r})$ by

$$\begin{aligned}
\varphi_n(\mathbf{r}) &= \phi_n(\mathbf{r}) + \int d\mathbf{r}' G_0(\mathbf{r}, \mathbf{r}', E)V_n(\mathbf{r}' - \mathbf{R}_n)\varphi(\mathbf{r}') \\
&= \phi_n(\mathbf{r}) + \int \int d\mathbf{r}'d\mathbf{r}'' G_0(\mathbf{r}, \mathbf{r}', E)t_n(\mathbf{r}', \mathbf{r}'', E)\phi_n(\mathbf{r}''),
\end{aligned} \quad (2.50)$$

which is the so-called Lippmann–Schwinger equation. We point out that both this relation and the Dyson equation for the t-matrix can be shown by applying the operator $(E - \hat{\mathcal{H}}_0(\mathbf{r}))$ in both sides of the equations themselves. The way to proceed now is to note that in the interstitial region of constant potential the wave function can be expressed in terms of plane waves. This allows to exploit the spherical symmetry of the problem by writing $\phi_n(\mathbf{r})$ as the following expansion

$$\phi_n(\mathbf{r}) \to \exp(i\mathbf{k} \cdot \mathbf{r}) = 4\pi \sum_L i^l j_l(kr)Y_L^*(\hat{r})Y_L(\hat{k}), \quad (2.51)$$

where the sum is over the pair of quantum numbers $L = (l, m)$, $Y_L(\hat{r})$ are the corresponding spherical harmonics,[2] and $j_l(kr)$ is the spherical Bessel function [18]. The single-particle Green's function can be expanded as a linear combination of spherical harmonics and Bessel functions too [18],

$$G_0(\mathbf{r}, \mathbf{r}', E) = -\frac{\exp(ik|\mathbf{r} - \mathbf{r}'|)}{4\pi|\mathbf{r} - \mathbf{r}'|} = -ik \sum_L j_l(kr_<)h_l^+(kr_>)Y_L(\hat{r})Y_L^*(\hat{r}'). \quad (2.52)$$

Here $r_> = \max\{r, r'\}$, $r_< = \min\{r, r'\}$, and $h_l^+(kr_>)$ are the spherical Hankel functions. Introducing Eqs. (2.51) and (2.52) into Eq. (2.50) and setting $r_n > r_{\text{MT},n}$ gives

$$\varphi_n(\mathbf{r}) = 4\pi \sum_L i^l Y_L^*(\hat{k}) \left[j_l(kr)Y_L(\hat{r}) - ik \sum_{L'} h_{l'}^+(kr)Y_{L'}(\hat{r})t_{n,L'L}(E) \right], \quad (2.53)$$

which defines the t-matrix in the angular momentum representation

$$t_{n,L'L}(E) = \int \int d\mathbf{r}d\mathbf{r}' j_{l'}(kr')Y_{L'}^*(\hat{r}')t_n(\mathbf{r}', \mathbf{r}, E)j_l(kr)Y_L(\hat{r}). \quad (2.54)$$

The solution presented in Eq. (2.53) is regular at the origin ($|\mathbf{r}_n| \to 0$) due to the good behaviour of the spherical Bessel functions employed. A different definition of

---

[2]They satisfy $\hat{L}^2 Y_L(\hat{r}) = l(l+1)Y_L(\hat{r})$ and $\hat{L}_z Y_L(\hat{r}) = mY_L(\hat{r})$, where $\hat{L}$ is the angular momentum operator.

regular solutions is commonly used too, which are in general known as the scattering solutions,

$$Z_{n,L}(\mathbf{r}, E) = \sum_{L'} j_{l'}(kr)Y_{L'}(\hat{r})t^{-1}_{n,L'L}(E) - ikh^+_l(kr)Y_L(\hat{r}), \quad \text{with } r_n > r_{\text{MT},n}.$$

(2.55)

In principle, irregular solutions should be included to fully describe the wave function too [18]. They are typically chosen as [18]

$$H_{n,L}(\mathbf{r}, E) = -ikh^+_l(kr)Y_L(\hat{r}).$$

(2.56)

These solutions must be matched to the form inside the bounding sphere, which can be obtained from Eq. (2.50) at $r_n < r_{\text{MT},n}$. Alternatively the differential form expressed in Eq. (2.45) can be used too, leading to the usual radial equation reflecting the spherical symmetry of the system. This operation determines the t-matrix and, therefore, yields the single-site problem solved. For computational purposes the angular momentum sums must be truncated at some maximum value. For most of the typical calculations with $l_{\text{max}} = 3$ give satisfactory results in terms of accuracy and at the same time yield computational affordable costs.

Finally, the single-site Green's function can be also obtained from the scattering solutions as [16, 18]

$$G_n(\mathbf{r}, \mathbf{r}', E) = \sum_{LL'} Z_{n,L}(\mathbf{r}, E)t_{n,LL'}(E)Z^\times_{n,L'}(\mathbf{r}, E) - \sum_L Z_{n,L}(\mathbf{r}_<, E)J^\times_{n,L}(\mathbf{r}_>, E),$$

(2.57)

where the superscript $^\times$ applies the conjugate operation on the spherical harmonics and $J_{n,L}(\mathbf{r}, E) = R_{n,L}(\mathbf{r}, E) - ikh^+_l(kr)Y_L(\hat{r})$, $R_{n,L}(\mathbf{r}, E)$ being the radial part of the wave solution.

2.2.2 Scattering Paths and the Multi-site Solution

The central idea of the MTA is to approximate the effective potential as a collection of non-overlapping single-centre scatterers. This construction has a remarkable consequence in the context of MST: the solution of the Kohn–Sham Hamiltonian, composed by the single-site potentials

$$\hat{\mathcal{H}}_{\text{KS}}(\mathbf{r}) = \hat{\mathcal{H}}_0(\mathbf{r}) + \sum_n V_n(r_n),$$

(2.58)

is completely determined by the scattering t-matrices associated with each scattering centre in their angular momentum representation, and the geometry of the problem. The starting point to show this is to construct the full T-matrix of the entire solid potential. Keeping $\hat{\mathcal{H}}_0$ as the non-perturbed reference system, the corresponding

Dyson equation is structured from every single-centre potential as well as the free-particle Green's function. This leads to the generalisation of Eq. (2.49) as

$$\underline{T} = \sum_n \underline{V}_n + \sum_n \underline{V}_n \underline{G}_0 \underline{T}. \tag{2.59}$$

We have switched to the angular momentum representation, which naturally yields matrix equations with angular momentum labels. Henceforth underlined quantities are used to represent their angular momentum matrix form, following the definition given in Eq. (2.53). For example, the components of $\underline{V}_n$ are

$$V_{n,LL'}(E) = \int \int \mathrm{d}\mathbf{r}\mathrm{d}\mathbf{r}' \, j_{l'}(kr') Y_{L'}^*(\hat{r}') V_n(\mathbf{r} - \mathbf{R}_n)\delta(\mathbf{r} - \mathbf{r}') j_l(kr) Y_L(\hat{r}). \tag{2.60}$$

Equation (2.59) defines the T-matrix in terms of itself. We can now expand the right hand side of this equation and express $\underline{T}$ as a sum involving the single-site potentials and $\underline{G}_0$. Note that now $\underline{G}_0$ connects potentials that are centred at different fixed positions. To continue exploiting the spherical symmetry of each scattering event, we expand $\underline{G}_0$ around two different centre positions. After carrying out the algebra one obtains [18]

$$\underline{G}_{0,nm,LL'}(E) = -4\pi i k \sum_{L_1} i^{l-l'-l_1} C_{LL'}^{L_1} Y_{L_1}(\hat{R}_{nm}) h_{l_1}^+(R_{nm}, E), \tag{2.61}$$

where $\mathbf{R}_{nm} = \mathbf{R}_m - \mathbf{R}_n$, $R_{nm} = |\mathbf{R}_{nm}|$, and the constants

$$C_{LL'}^{L_1} = \int \mathrm{d}\hat{r} \, Y_L(\hat{r}) Y_{L'}^*(\hat{r}) Y_{L_1}(\hat{r}) \tag{2.62}$$

are known as the Gaunt numbers. By repeating substitution of $\underline{T}$ into Eq. (2.59), and using the Dyson equation of the t-matrix in the angular momentum representation ($\underline{t}_n = \underline{V}_n + \underline{V}_n \underline{G}_0 \underline{t}_n$), one can finally write the central multiple scattering equation

$$\begin{aligned}
\underline{T} &= \sum_n \underline{V}_n + \sum_{nm} \underline{V}_n \underline{G}_{0,nm} \underline{V}_m + \sum_{nmk} \underline{V}_n \underline{G}_{0,nm} \underline{V}_m \underline{G}_{0,mk} \underline{V}_k + \cdots \\
&= \sum_{nm} \left[\underline{t}_n \delta_{nm} + \underline{t}_n \underline{G}_{0,nm}(1 - \delta_{nm})\underline{t}_m + \cdots \right].
\end{aligned} \tag{2.63}$$

Hence, $\underline{T}$ depends on the matrices $\{\underline{t}_n\}$ and $\underline{G}_{0,mn}$, as we were seeking to show. From this result we can introduce the scattering path operator (SPO), $\underline{\tau}_{mn}$,

$$\underline{T} = \sum_{mn} \underline{\tau}_{mn} \Rightarrow \underline{\tau}_{mn} = \underline{t}_n \delta_{nm} + \underline{t}_n \sum_k \underline{G}_{0,nk}(1 - \delta_{nk})\underline{\tau}_{km}. \tag{2.64}$$

The SPO gives a useful physical interpretation of the multiple scattering theory presented here. By inspecting the effect of applying the operator $\underline{G}_0 \underline{\tau}_{mn}$ on the incoming free-electron plane wave, and recalling the Lippman–Schwinger equation, it is clear that the SPO sums the effect of all scattering paths beginning at site n and finishing at site m. The T-matrix then comprises every possible scattering event in the entire solid. The limitation of Eq. (2.63), as stated at the beginning of the section, is that the potentials must not overlap. This is manifest by recalling that the t-matrices are obtained by matching the spherical solutions to the free-particle plane wave at the muffin-tin radii. Overlapping potentials break down this spatial pattern so that the calculation would be inconsistent. In addition to this, as pointed out in Refs. [15, 17], the form given in Eq. (2.61) should not be used at the superposed regions, which yields the real space version of Eq. (2.63) intractable for overlapping potentials. Despite this, the ASA is frequently used.

A more convenient expression for the SPO can be directly obtained by manipulating Eq. (2.64). The following objects whose components are matrices in the angular momentum representation must be defined first,

$$\tau(E) = \{\underline{\tau}_{nm}\}, \quad t(E) = \{\underline{t}_n \delta_{nm}\}, \quad G_0(E) = \{(1 - \delta_{nm})\underline{G}_{0,nm}\}. \tag{2.65}$$

Introducing this into Eq. (2.64) gives

$$\tau(E) = \left(t^{-1}(E) - G_0(E)\right)^{-1}. \tag{2.66}$$

This is another central equation that will show itself as very useful in future derivations and manipulations.

To conclude this section, we finally show an expression for the full Green's function associated with the multiple scattering problem. By using the scattering solutions and the corresponding Dyson equation one can finally write [18]

$$\begin{aligned} G(\mathbf{r}, \mathbf{r}', E) = {} & \underline{Z}_n(\mathbf{r} - \mathbf{R}_n, E)\underline{\tau}_{nm}(E)\underline{Z}_m^\times(\mathbf{r}' - \mathbf{R}_m, E) \\ & - \delta_{nm}\underline{Z}_n(\mathbf{r}_< - \mathbf{R}_n, E)\underline{J}_n^\times(\mathbf{r}_> - \mathbf{R}_n, E), \end{aligned} \tag{2.67}$$

where $\underline{Z}_n(\mathbf{r}, E)$ and $\underline{J}_n(\mathbf{r}, E)$ are vectors with components labelled by angular momentum indices. The index n (m) in the right hand side is chosen such that the vector $\mathbf{r}$ ($\mathbf{r}'$) is inside the respective spherical potential. The KKR-MST strategy is clear now. The SPO is constructed via application of Eq. (2.66). It is composed by the t-matrices containing all atomic-like information and the scatterer connector $G_{0,nm}$, which is determined separately by the geometry of the system. Once this result is obtained the T-matrix and the Green's function can be directly calculated from Eqs. (2.64) and (2.67), respectively.

2.2.3 *Density Calculation and the Lloyd Formula*

This section is devoted to show how to use $G(\mathbf{r}, \mathbf{r}', E)$ to obtain the electron and magnetic moment densities. We start by expressing the formal definition of the Green's function given in Eq. (2.47) in terms of the eigenvalues and eigenfunctions of the Hamiltonian of interest. Naming these quantities $\{E_i\}$ and $\{\psi_i(\mathbf{r})\}$ for $\hat{\mathcal{H}}_{\mathrm{KS}}(\mathbf{r})$, respectively, one can write

$$G^{\pm}(\mathbf{r}, \mathbf{r}', E) = \lim_{\epsilon \to \pm 0} \sum_i \frac{\psi_i(\mathbf{r})\psi_i^*(\mathbf{r}')}{E - E_i + i\epsilon} = \sum_i \psi_i(\mathbf{r})\psi_i^*(\mathbf{r}')\left(\mathcal{P}\frac{1}{E - E_i} \mp i\pi\delta(E - E_i)\right),$$

$$(2.68)$$

where $\mathcal{P}$ stands for the Cauchy principal value. By recalling Eq. (2.39), it directly follows that

$$n(\mathbf{r}) = \mp\frac{1}{\pi}\mathrm{Im}\,\mathrm{Tr}\int \mathrm{d}E f(E) G^{\pm}(\mathbf{r}, \mathbf{r}, E), \tag{2.69}$$

where $f(E)$ is the appropriate occupation function and $\mathrm{Tr}[\ldots]$ traces over spin coordinates for the two-spinor Dirac solution in SDFT if necessary. Similarly, the magnetic moment density is

$$\boldsymbol{\mu}(\mathbf{r}) = \pm\mu_{\mathrm{B}}\frac{1}{\pi}\mathrm{Im}\,\mathrm{Tr}\int \mathrm{d}E f(E)\beta_I \boldsymbol{\Sigma} G^{\pm}(\mathbf{r}, \mathbf{r}, E), \tag{2.70}$$

where

$$\beta_I \boldsymbol{\Sigma} = \beta_I \begin{pmatrix} \boldsymbol{\sigma} & (0) \\ (0) & \boldsymbol{\sigma} \end{pmatrix} = \begin{pmatrix} \boldsymbol{\sigma} & (0) \\ (0) & -\boldsymbol{\sigma} \end{pmatrix}. \tag{2.71}$$

The density of states at energy E can be provided by the Green's function too after performing similar manipulations

$$n(E) \equiv \sum_i \delta(E - E_i) = \mp\frac{1}{\pi}\mathrm{Im}\,\mathrm{Tr}\int \mathrm{d}\mathbf{r}\, G^{\pm}(\mathbf{r}, \mathbf{r}, E) = \mp\frac{1}{\pi}\mathrm{Im}\,\mathrm{Tr}\, G^{\pm}(E). \tag{2.72}$$

Note that the definition given in Eq. (2.46) has been employed to derive the last equality. The total number of electrons is then

$$N = \int_{-\infty}^{\infty} \mathrm{d}E f(E) n(E) = \mp\frac{1}{\pi}\mathrm{Im}\,\mathrm{Tr}\int_{-\infty}^{\infty} \mathrm{d}E f(E)\int \mathrm{d}\mathbf{r}\, G^{\pm}(\mathbf{r}, \mathbf{r}, E), \tag{2.73}$$

which becomes

$$N(E_{\mathrm{F}}) = N_0(E_{\mathrm{F}}) + \delta N(E_{\mathrm{F}}), \tag{2.74}$$

with

$$N_0(E_{\mathrm{F}}) = \pm\frac{1}{\pi}\mathrm{Im}\,\mathrm{Tr}\log G_0^{\pm}(E_{\mathrm{F}}), \tag{2.75}$$

$$\delta N(E_{\mathrm{F}}) = \pm \frac{1}{\pi} \mathrm{ImTr} \log T^{\pm}(E_{\mathrm{F}}), \tag{2.76}$$

when $f(E)$ adopts the shape of the Fermi–Dirac function in the limit of approaching zero temperature, i.e. $f(E < E_{\mathrm{F}}) = 1$ and $f(E > E_{\mathrm{F}}) = 0$, for E_{F} being the Fermi energy. The serviceable Dyson equation as well as some identities of the Green's function and the T-matrix have been used to arrive to Eqs. (2.75) and (2.76) [18], which are known as the famous Lloyd formula.

2.3 The Effective Medium Theory of Disorder: The Coherent Potential Approximation

This thesis focuses on the study of magnetism at finite temperatures. Particularly, our theory aims to tackle the problem of thermally fluctuating disordered local moments at every magnetic site and its effect on the electronic structure. A scheme to perform the appropriate ensemble averages over the corresponding magnetic degrees of freedom is, therefore, required to find the equilibrium properties of the system. Fortunately, a method to approximate a potential containing site-dependent disorder, and suitable for the particularities of our theory (see Sect. 3.3), was developed a few decades ago within the context of MST, the coherent potential approximation (CPA) [15, 20]. In spite of the fact that the CPA was initially motivated to describe substitutional disordered alloys, it can be used for other types of disorder, such as magnetic disorder as pertinent to our work [21]. It was recognised early that the adequate object of study is the averaged Green's function. Following the work of many other authors, P. Soven finally presented the central idea of the approximation in the paper which is nowadays considered as the basic reference of CPA [20]. Although this early formulation was already based on the most basic objects of MST, namely the Green's function and the t-matrix, it was not until the work of Győrffy and co-authors that the CPA was expressed in terms of the SPO in the angular momentum representation for a muffin-tin approximated potential, hence yielding a numerically tractable scheme [15, 22].

The CPA consists in replacing the disordered crystal potential by an effective ordered medium that forces the electrons to adopt the average behaviour as closely as possible [15, 20, 22–24]. The key idea that made the CPA successful compared with its rival alternative, the averaged t-matrix approximation (ATA) [25], was that the CPA is shaped as a self-consistent approach. While in the ATA the effective potential is chosen beforehand, usually following the virtual crystal model, in the CPA it plays the role of a parameter that is found by satisfying some more sophisticated criterion. By testing the predictions of both approximations the CPA has been shown to give more accurate results compared with the ATA. In fact, the CPA is usually referred to as the best possible single-site approximation [23].

The CPA condition is that the scattering occurring if the disordered site is embedded by the effective medium is exactly the same as the generated if the site is filled by the effective medium itself. Returning to scattering theory language, the effective medium is mapped to the so-called coherent potential, $V_c(\mathbf{r}) = \sum_n V_n(|\mathbf{r} - \mathbf{R}_n|)$, constructed with a muffin-tin shape, as defined in Eq. (2.44). As usual, the Dyson equation can be invoked to write the central scattering quantities at each site of the coherent lattice in the angular momentum representation,

$$\underline{G}_{c,n} = \underline{G}_0 + \underline{G}_0 \underline{t}_{c,n} \underline{G}_0, \tag{2.77}$$

$$\Rightarrow \underline{t}_{c,n} = \underline{V}_{c,n} + \underline{V}_{c,n} \underline{G}_0 \underline{t}_{c,n}. \tag{2.78}$$

Here $\underline{G}_0$ is the Green's function of the free-particle, and $\underline{G}_{c,n}$ and $\underline{t}_{c,n}$ are the Green's function and t-matrix associated with the coherent potential at site n, respectively. Let's assume now that the disorder at a chosen site n_0 is determined by a combination of different single-site potentials $\{V_{\alpha,n_0}(|\mathbf{r} - \mathbf{R}_{n_0}|)\}$, enumerated by α and with probability P_{α,n_0} to randomly exist at the site.[3] The effect of replacing the muffin-tin sphere of the coherent potential by $V_{\alpha,n_0}(|\mathbf{r} - \mathbf{R}_{n_0}|)$ at site n_0 can be described by

$$\underline{G}_{\alpha,n_0} = \underline{G}_{c,n_0} + \underline{G}_{c,n_0} \underline{t}_{\alpha,n_0}^{\mathrm{diff}} \underline{G}_{c,n_0}, \tag{2.79}$$

where $\underline{t}_{\alpha,n_0}^{\mathrm{diff}}$ is not the appropriate t-matrix of $V_{\alpha,n_0}(|\mathbf{r} - \mathbf{R}_{n_0}|)$, but one that accounts for the difference between this and the coherent potential, that is

$$\underline{t}_{\alpha,n_0}^{\mathrm{diff}} = \left(\underline{V}_{\alpha,n_0} - \underline{V}_{c,n_0} \right) + \left(\underline{V}_{\alpha,n_0} - \underline{V}_{c,n_0} \right) \underline{G}_{c,n_0} \underline{t}_{\alpha,n_0}^{\mathrm{diff}}. \tag{2.80}$$

Clearly, from these relations the CPA condition can be translated to the scattering terminology as

$$\sum_\alpha P_{\alpha,n_0} \underline{t}_{\alpha,n_0}^{\mathrm{diff}} = 0, \tag{2.81}$$

which evidences the single-site nature of the approximation. Note that this equation can be expressed in terms of Green's functions from Eq. (2.79) as

$$\underline{G}_{c,n_0} = \sum_\alpha P_{\alpha,n_0} \underline{G}_{\alpha,n_0}. \tag{2.82}$$

In order to make the CPA implementable, it is desirable to write Eq. (2.81) in terms of the SPO. To this end, we consider separately each impurity problem with probability P_{α,n_0}, i.e. a lattice where the coherent potential is present at every site apart from n_0, which is filled by $V_{\alpha,n_0}(\mathbf{r}_{n_0})$. If we write $\underline{T}_{\alpha(n_0)}$ as the appropriate multiple scattering T-matrix in this situation, the Dyson equation for the total Green's function is

$$\underline{G}_{\alpha(n_0)} = \underline{G}_0 + \underline{G}_0 \underline{T}_{\alpha(n_0)} \underline{G}_0, \tag{2.83}$$

---

[3]Evidently, the probabilities are restricted to satisfy $\sum_\alpha P_{\alpha,n_0} = 1$.

where the subscript (n_0) is added to indicate the site of the impurity. This expression together with Eqs. (2.81) and (2.82) allows to connect $\{\underline{T}_{\alpha(n_0)}\}$ to the total scattering T-matrix of the coherent potential lattice $\underline{T}_c$ as

$$\underline{T}_c = \sum_\alpha P_{\alpha,n_0} \underline{T}_{\alpha(n_0)}, \tag{2.84}$$

which in turn may be written in terms of the SPO by making use of Eq. (2.64) [15],

$$\underline{\tau}_{c,n_0 n_0} = \sum_\alpha P_{\alpha,n_0} \underline{\tau}_{\alpha(n_0),n_0 n_0}. \tag{2.85}$$

The last two expressions show that the effective medium can be also understood as a fictitious system that produces a multiple scattering effect that is equal to the corresponding average at the site, when it is indeed surrounded by the coherent medium itself. The CPA is, therefore, referred to as the mean-field approximation in scattering theory. We point out that Eq. (2.85) has been able to be spatially resolved thanks to the fact that the switch to the angular momentum representation can be conveniently realised for non-overlapping muffin-tin spheres [15]. Moreover, a solution for the impurity SPO can be obtained by recalling Eq. (2.66). Considering the (n_0, n_0)-component of this equation only gives

$$\underline{\tau}_{c,n_0 n_0}^{-1} = \underline{\tau}_{\alpha(n_0),n_0 n_0}^{-1} - \underline{t}_{\alpha(n_0),n_0}^{-1} + \underline{t}_{c,n_0}^{-1}, \tag{2.86}$$

where now $\underline{t}_{\alpha(n_0),n_0}^{-1}$ is the appropriate t-matrix at the disordered site of the impurity problem. After some trivial manipulation of Eq. (2.86) one can write [22]

$$\underline{\tau}_{\alpha(n_0),n_0 n_0} = \underline{D}_{\alpha,n_0} \underline{\tau}_{c,n_0 n_0}, \tag{2.87}$$

where

$$\underline{D}_{\alpha,n_0} = \left[\underline{1} + \left(\underline{t}_{\alpha(n_0),n_0}^{-1} - \underline{t}_{c,n_0}^{-1} \right) \underline{\tau}_{c,n_0 n_0} \right]^{-1} \tag{2.88}$$

is the so-called impurity matrix. Similarly, one can define the following object, known as the excess scattering matrix,

$$\underline{X}_{\alpha,n_0} = \left[\left(\underline{t}_{\alpha(n_0),n_0}^{-1} - \underline{t}_{c,n_0}^{-1} \right)^{-1} + \underline{\tau}_{c,n_0 n_0} \right]^{-1}, \tag{2.89}$$

which satisfies

$$\underline{\tau}_{\alpha(n_0),n_0 n_0} = \underline{\tau}_{c,n_0 n_0} + \underline{\tau}_{c,n_0 n_0} \underline{X}_{\alpha,n_0} \underline{\tau}_{c,n_0 n_0}, \tag{2.90}$$

since $\underline{D}_{\alpha,n_0} = \underline{1} + \underline{\tau}_{c,n_0 n_0} \underline{X}_{\alpha,n_0}$. From Eq. (2.87) the CPA prescription given in Eq. (2.85) can be conveniently expressed as

$$\sum_{\alpha} P_{\alpha,n_0} \underline{D}_{\alpha,n_0} = \underline{1}, \tag{2.91}$$

or alternatively as

$$\sum_{\alpha} P_{\alpha,n_0} \underline{X}_{\alpha,n_0} = \underline{0}, \tag{2.92}$$

and the CPA can be finally numerically implemented by solving these equations self-consistently. The approach presented here allows to obtain $\{\underline{t}_{c,n_0}^{-1}\}$ and $\underline{\tau}_{c,n_0 n_0}$ from $\{\underline{t}_{\alpha(n_0),n_0}^{-1}\}$, which are the quantities naturally calculated from the scattering theory introduced earlier, and the structural information to construct the two-site expanded Green's functions. We would like to point out that although the solution of both Eqs. (2.91) and (2.92) corresponds to the same effective medium, the CPA implementation using the excess scattering matrix is preferable. As it was shown by R. Mills, an iterative scheme ensuring stable convergence to a unique self-consistent solution exists for Eq. (2.92) [26].

References

1. Martin R (2004) Electronic structure - basic theory and practical methods. CUP, Cambridge
2. Giustino F (2014) Materials modelling using density functional theory. Oxford University Press, Oxford
3. Hohenberg P, Kohn W (1964) Inhomogeneous electron gas. Phys Rev 136:B864–B871
4. Kohn W, Sham LJ (1965) Self-consistent equations including exchange and correlation effects. Phys Rev 140:A1133–A1138
5. Ceperley DM, Alder BJ (1980) Ground state of the electron gas by a stochastic method. Phys Rev Lett 45:566–569
6. Perdew JP, Zunger A (1981) Self-interaction correction to density-functional approximations for many-electron systems. Phys Rev B 23:5048–5079
7. Kittel C (1976) Introduction to solid state physics. Wiley, New York
8. Perdew John P, Wang Yue (1992) Accurate and simple analytic representation of the electron-gas correlation energy. Phys Rev B 45:13244–13249
9. Gunnarsson O, Lundqvist BI (1976) Exchange and correlation in atoms, molecules, and solids by the spin-density-functional formalism. Phys Rev B 13:4274–4298
10. David Mermin N (1965) Thermal properties of the inhomogeneous electron gas. Phys Rev 137:A1441–A1443
11. Rajagopal AK, Callaway J (1973) Inhomogeneous electron gas. Phys Rev B 7:1912–1919
12. Korringa J (1947) On the calculation of the energy of a Bloch wave in a metal. Physica 13(6):392–400
13. Kohn W, Rostoker N (1954) Solution of the Schrödinger equation in periodic lattices with an application to metallic lithium. Phys Rev 94:1111–1120
14. Lloyd P, Smith PV (1972) Multiple scattering theory in condensed materials. Adv Phys 21(89):69–142
15. Gyorffy BL (1972) Coherent-Potential Approximation for a nonoverlapping-muffin-tin-potential model of random substitutional alloys. Phys Rev B 5:2382–2384
16. Strange P (1998) Relativistic quantum mechanics. Cambridge University Press, Cambridge
17. Phariseau P, Györffy BL, Scheire L (1979) Electron in disordered metals and at metallic surfaces. Plenum Press, New York

18. Zabloudil J, Hammerling R, Szunyogh L, Weinberger P (2005) Electron scattering in solid matter. Springer, Berlin
19. Gonis A (1992) Green functions for ordered and disordered systems. North-Holland, Amsterdam
20. Soven Paul (1967) Coherent-potential model of substitutional disordered alloys. Phys Rev 156:809–813
21. Gyorffy BL, Pindor AJ, Staunton J, Stocks GM, Winter H (1985) A first-principles theory of ferromagnetic phase transitions in metals. J Phys F Met Phys 15(6):1337
22. Durham PJ, Gyorffy BL, Pindor AJ (1980) On the fundamental equations of the Korringa-Kohn-Rostoker (KKR) version of the coherent potential approximation (CPA). J Phys F Met Phys 10(4):661
23. Ehrenreich H, Schwartz LM (1976) The electronic structure of alloys. Solid state physics, vol 31. Academic, Cambridge, pp 149–286
24. Faulkner JS (1982) The modern theory of alloys. Prog Mater Sci 27(1):1–187
25. Korringa J (1958) Dispersion theory for electrons in a random lattice with applications to the electronic structure of alloys. J Phys Chem Solids 7(2):252–258
26. Mills R, Gray LJ, Kaplan T (1983) Analytic approximation for random muffin-tin alloys. Phys Rev B 27:3252–3262

Chapter 3
Disordered Local Moment Theory and Fast Electronic Responses

Solid systems present fascinating and intriguing phenomena whose physical origin is underlined by the complicated motions and interactions among the collections of electrons and nuclei. Chapter 2 has been devoted to introduce DFT as a powerful technique to tackle this many-body problem and describe the electronic structure. However, further complexity is gained if finite temperature effects are taken into account, leading to additional obstacles that can be even more difficult to overcome. The effect of temperature is to produce random deviations from the averaged state of relevant physical quantities, which become larger and more frequent as the temperature is increased. For example, fluctuations of the atomic positions and veloci-ties originate collective vibrational modes, or phonons, while electronic excitations within the electronic band structure can be thermally activated too. Moreover, when magnetic moments exist they can act cooperatively to form diverse magnetically-ordered structures. Fluctuations of the magnetic moment orientations and sizes are thermally induced such that phase transitions between very different magnetic phases can be triggered by temperature changes. The magnetic ordering is eventually van-ished at high enough temperatures. An outstanding challenge in its own right is to explain the temperature-dependent properties of large variety of magnetic phases and magnetic transitions from an ab initio perspective.

In this chapter we develop the part of the theory designed to describe thermal fluctuations of magnetic degrees of freedom. Firstly, in Sect. 3.1 we introduce the basic concepts and framework of magnetism at finite temperatures from which we base our theory. The following sections are devoted to show how SDFT, together with KKR-MST formalism, can be used as a natural technique to describe efficiently transverse excitations of fluctuating local magnetic moments within a disordered local moment picture. The theory can be divided into two central parts. One refers to the construction of a Hamiltonian for the evolution of the local moment orientations, given in Sect. 3.2. As will be shown, such part of the approach can then be incor-porated adequately within KKR-MST in terms of the serviceable Green's functions. This is the second part of the problem and it is treated in Sect. 3.3.1. Finally, Sect. 3.4

© Springer Nature Switzerland AG 2020
E. Mendive Tapia, *Ab initio Theory of Magnetic Ordering*, Springer Theses,
https://doi.org/10.1007/978-3-030-37238-5_3

is meant to show the limit of the completely disordered local moment state at high temperatures and how the lattice Fourier transform of some key quantities in this regime can provide insightful information of the type of magnetic ordering that can be stabilised.

3.1   Magnetism at Finite Temperatures and Conceptual Framework

As discussed a few decades ago in various seminal works [1–6], SDFT descriptions inevitably lead to the calculation of unrealistically high critical temperatures of ferromagnetic phase transitions. Although Mermin generalized DFT to finite temperatures [7], its computational implementation in the context of magnetic fluctuations has clear complications. Consider a magnetic system with magnetic moments whose orientations thermally fluctuate. The equilibrium properties at a given temperature are formally calculated by finding the charge and magnetic moment densities that suitably minimise the associated grand potential functional. Evidently, finding this object is extremely difficult and describing approximately magnetic moment densities describing the disordered local moment state within the LDA in principle would demand the construction of prohibitively large magnetic unit cells. Clearly, the problem presented here is an overwhelming task in both conceptual and computational aspects which cannot be overcome by brute force calculations. In practice, traditional SDFT is designed to calculate magnetic states that posses the periodicity of the magnetic unit cell and, therefore, transverse spin fluctuations are naturally ignored. Hence, the state with zero magnetisation is reached only when the size of the magnetic moment collapses due to Stoner excitations between up and down spin bands. This process requires the application of very high temperatures because of the magnitude of the exchange splitting size [3, 8], leading to the overestimation of the Curie temperature of a ferromagnetic phase transition, for example. Moreover, reasonable transition temperatures should be expected to be obtained when the orientations of the magnetic moments are allowed to thermally fluctuate, and hence describing rigorously the paramagnetic limit. In this situation the total magnetisation vanishes as a result of the formation of fully disordered local moments, instead of the reduction of the magnetic moment length [5, 6], which requires much lower energy excitations.

Pertinent to this observation is the early view of itinerant magnetism and indirect exchange interaction adopted by many authors in the past [2, 3, 5]. They considered the existence of magnetic moments to be owed to sufficiently localised exchange interactions mediated by itinerant electrons hopping among magnetic sites, and related it to the notion of local magnetic fields (or exchange fields). These fields, sustaining the magnetic moments, emerge from the motions and interactions among the electrons and nuclei and are localised at each magnetic atom. The process is self-consistent in nature: the exchange fields directly affect the spin of the travelling electrons whose interactions, in turn, give rise to the local fields themselves. This picture is often interpreted as some fictitious magnetic glue that arises from the spin-polarised electronic structure. Evidently, the Coulomb pair repulsion and the Pauli exclusion principle are fundamentally behind the origin of this effect. In principle, SDFT can be used to evaluate the strength of the local magnetic fields [9]. However, in the disordered local moment state they change their direction from atom to atom accordingly to the local moment directions. Clearly, their calculation for systems with magnetic disorder is complicated and the treatment of magnetic fluctuations still faces the problem of finding the appropriate grand potential functional. Nonetheless, by incorporating further elaborations along this route in the context of KKR-MST, in 1985 Győrffy et al. [6] presented an ab initio theory appropriate for local magnetic moments that are randomly disordered, referred to as Disordered Local Moment (DLM) theory. In this thesis we use this approach and extend it.

DLM theory is designed to calculate the local magnetic fields, emerging from the electronic structure underlying the corresponding magnetic state, by exploiting SDFT machinery. The crucial point is that the mathematical features of the theory naturally allow to construct an effective medium to describe disordered magnetic configurations and, hence, realistic temperature-dependent properties can be accomplished by affordable calculations. The essential idea behind this is based on the comparison of time scales between different electronic degrees of freedom [5, 6]. Consider again the process of magnetic moment formation to be described by spin-correlated electronic interactions within the fast time scale of electron hopping, τ_{elec}. The orientations of these moments, emerging from the rapidly evolving electronic surroundings, are then assumed to be slowly varying degrees of freedom that remain unchanging in a time scale of $\tau_{\text{form}} > \tau_{\text{elec}}$. Time averages over τ_{form}, therefore, render the system confined to a phase space prescribed by a collection of unitary vectors, $\{\hat{e}_n\}$, that specify the orientations of the local moments at each magnetic site n. The ergodicity is, in consequence, temporarily broken. The key assumption is that τ_{form} is short compared with the time $\ll_{\text{wave}}$ necessary for the local moments to change their orientations, which is in principle attributed to the time scale of the occurrence of spin wave excitations. Note that this approach is essentially the adiabatic approximation of slowly varying local moments. The fast electronic motions evolve with a background of frozen local moments whose gradual change is governed by the energy of each magnetic configuration. The directions $\{\hat{e}_n\}$, which play somewhat an analogous role compared with the fixed nuclei positions in the Born–Oppenheimer approximation, can be treated as independent parameters at short time scales and the theory, consequently, is aimed to adequately evaluate the response of the electronic structure to magnetic configurations described by $\{\hat{e}_n\}$.

We should argue now that systems whose magnetism is described well by the picture proposed above allow to approximate the trace operation in Eqs. (2.16), (2.17), and (2.18) as

$$\text{Tr} \rightarrow \text{Tr}_{\{\hat{e}_n\}} \text{Tr}_{\text{rest}}, \tag{3.1}$$

where the trace over the orientations $\{\hat{e}_n\}$ is explicitly separated from the trace over the rest of the electronic degrees of freedom. If we proceed by minimising the grand potential functional introduced in Eq. (2.16) by the appropriate probability density,[1]

$$\rho_0 = \frac{\exp\left[-\beta(\hat{\mathcal{H}} - \nu\hat{N})\right]}{\mathrm{Tr}_{\{\hat{e}_n\}}\mathrm{Tr}_{\mathrm{rest}}\exp\left[-\beta(\hat{\mathcal{H}} - \nu\hat{N})\right]},$$ (3.2)

the grand potential can be written as

$$\Omega = -\frac{1}{\beta}\log\left(\mathrm{Tr}_{\{\hat{e}_n\}}\mathrm{Tr}_{\mathrm{rest}}\exp\left[-\beta(\hat{\mathcal{H}} - \nu\hat{N})\right]\right),$$ (3.3)

$$= -\frac{1}{\beta}\log\left(\mathrm{Tr}_{\{\hat{e}_n\}}\exp\left[-\beta\tilde{\Omega}(\{\hat{e}_n\})\right]\right),$$ (3.4)

which formally defines a grand potential constrained to the magnetic configuration $\{\hat{e}_n\}$,

$$\tilde{\Omega}(\{\hat{e}_n\}) = -\frac{1}{\beta}\log\left(\mathrm{Tr}_{\mathrm{rest}}\exp\left[-\beta(\hat{\mathcal{H}} - \nu\hat{N})\right]\right).$$ (3.5)

We should emphasise that the directions $\{\hat{e}_n\}$ are classical quantities emerging from the many electron interacting system and restricting the orientations of the local moments resulting from the problem set by Eq. (3.5). In the context of SDFT, this means that at short time scales τ_{form} the appropriate electron and magnetic moment densities, $n(\mathbf{r})$ and $\mu(\mathbf{r})$, are forced to satisfy

$$\mu_n\hat{e}_n = \int_{V_n}\mathrm{d}\mathbf{r}\,\mu(\mathbf{r}).$$ (3.6)

Equation (3.6) states that $\mu(\mathbf{r})$ is such that inside every region of volume V_n centred at atomic site n there is a total spin polarisation with an orientation constrained to be along $\hat{e}_n$. Both densities $n(\mathbf{r})$ and $\mu(\mathbf{r})$ minimise, therefore, the constrained grand potential functional

$$\tilde{\Omega}[\hat{\rho}_{\mathrm{rest}}] = \mathrm{Tr}_{\mathrm{rest}}\hat{\rho}_{\mathrm{rest}}\left(\hat{\mathcal{H}} - \nu\hat{N} + \frac{1}{\beta}\log\hat{\rho}_{\mathrm{rest}}\right),$$ (3.7)

for the probability density

$$\hat{\rho}_{\mathrm{rest},0} = \frac{\exp\left[-\beta\left(\hat{\mathcal{H}} - \nu\hat{N}\right)\right]}{\mathrm{Tr}_{\mathrm{rest}}\exp\left[-\beta\left(\hat{\mathcal{H}} - \nu\hat{N}\right)\right]}.$$ (3.8)

---

[1]We remind the reader that $\hat{\mathcal{H}}$, ν, and $\hat{N}$ are the many-body Hamiltonian, the chemical potential, and the particle number operator of the entire solid problem, respectively.

Evidently, the equilibrium magnetic properties of the system are calculated by carrying out the ensemble averages over all local moment orientational configurations $\{\hat{e}_n\}$. Note that due to the continuous nature of $\{\hat{e}_n\}$, the trace $\text{Tr}_{\{\hat{e}_n\}}$ is applied by performing the integrals

$$\text{Tr}_{\{\hat{e}_n\}} \rightarrow \prod_n \int d\hat{e}_n. \tag{3.9}$$

Thus, in the equilibrium every magnetic configuration is appropriately weighted by the probability

$$P(\{\hat{e}_n\}) = \frac{\exp\left[-\beta\tilde{\Omega}(\{\hat{e}_n\})\right]}{\prod_{n'} \int d\hat{e}_{n'} \, \exp\left[-\beta\tilde{\Omega}(\{\hat{e}_n\})\right]}. \tag{3.10}$$

The grand potential, in consequence, is obtained by performing the following average

$$\Omega = \prod_{n'} \int d\hat{e}_{n'} P(\{\hat{e}_n\}) \left(\tilde{\Omega}(\{\hat{e}_n\}) + \frac{1}{\beta} \log P(\{\hat{e}_n\}) \right). \tag{3.11}$$

From the view of Eqs. (3.5) and (3.11), $\tilde{\Omega}(\{\hat{e}_n\})$ can be interpreted as a Hamiltonian governing the orientations of the local moments, which is characterised by the behaviour of the rapidly responsive electronic structure and can be specified for every orientational configuration $\{\hat{e}_n\}$. We remark that the electronic origin of this object makes its dependence on $\{\hat{e}_n\}$ expectedly very complicated in metals, where the magnetic interactions are strongly mediated by itinerant electrons. Instead of working with it, the strategy is to construct a simpler trial Hamiltonian $\mathcal{H}_0(\{\hat{e}_n\})$ and invoke the Peierls–Feynman inequality to find an upper bound of Ω [10, 11]. We structure the DLM scheme as follows

1. *Evolution of the local moment orientations:* The first part consists in constructing $\mathcal{H}_0(\{\hat{e}_n\})$ and perform the corresponding statistical mechanics to describe the behaviour of the local moments at long time scales.
2. *Electronic structure part:* The second part refers to the description at short time scales τ_{form}. The aim is to extract as best as possible the dependence of $\tilde{\Omega}(\{\hat{e}_n\})$ on the local moment directions by performing appropriate SDFT calculations constrained to magnetic moment densities described by $\{\hat{e}_n\}$. It is the KKR-MST formalism presented in Sect. 2.2 the pertinent technology for this task.

We firstly proceed by constructing $\mathcal{H}_0(\{\hat{e}_n\})$ and present a mean-field theory for the description of the local moment orientations in Sect. 3.2, which will be followed by Sect. 3.3, devoted to the electronic structure part of the problem. We advance that KKR-MST formalism allows to naturally perform the statistical averages over the orientations $\{\hat{e}_n\}$ thanks to the adequate implementation of the CPA within the mean-field approximation of $\mathcal{H}_0$. Thereby, DLM theory is designed to evaluate the ensemble average of $\tilde{\Omega}(\{\hat{e}_n\})$, consistent with the effective medium, directly from SDFT calculations.

3.2 Mean-Field Theory and Statistical Mechanics of Disordered Local Moments

In Sect. 3.1 we have explained that in the DLM approach the motions of the local moment orientations are described by $\tilde{\Omega}(\{\hat{e}_n\})$. This is the Hamiltonian that accurately carries the complexity from the mixing of delocalised and more local interacting electrons, extracted from SDFT calculations. In other words, it contains the interactions among the local moments and their coupling with itinerant electron spin effects. Contrary to magnetic insulators, in which the localised nature of the electrons and their spins usually allows for a description of their magnetism in terms of a simple Heisenberg type pairwise Hamiltonian, for metallic systems the magnetic interactions are in principle more complicated and, consequently, $\tilde{\Omega}(\{\hat{e}_n\})$ might contain multi-spin, or multi-site, interactions. This means that for example

$$\tilde{\Omega}(\dots \uparrow \uparrow \uparrow \uparrow \uparrow \uparrow \uparrow \uparrow \uparrow \uparrow \dots) + \tilde{\Omega}(\dots \uparrow \uparrow \downarrow \uparrow \uparrow \uparrow \downarrow \uparrow \uparrow \uparrow \dots)$$
$$-\tilde{\Omega}(\dots \uparrow \uparrow \downarrow \uparrow \uparrow \uparrow \uparrow \uparrow \uparrow \uparrow \dots) - \tilde{\Omega}(\dots \uparrow \uparrow \uparrow \uparrow \uparrow \uparrow \downarrow \uparrow \uparrow \uparrow \dots),$$

$$\tag{3.12}$$

which is a calculation designed to yield the two-site interaction, gives a different value than

$$\tilde{\Omega}(\dots \uparrow \downarrow \uparrow \downarrow \uparrow \downarrow \uparrow \downarrow \uparrow \downarrow \dots) + \tilde{\Omega}(\dots \uparrow \downarrow \downarrow \downarrow \uparrow \downarrow \downarrow \downarrow \uparrow \downarrow \dots)$$
$$-\tilde{\Omega}(\dots \uparrow \downarrow \downarrow \downarrow \uparrow \downarrow \uparrow \downarrow \uparrow \downarrow \dots) - \tilde{\Omega}(\dots \uparrow \downarrow \uparrow \downarrow \uparrow \downarrow \downarrow \downarrow \uparrow \downarrow \dots).$$

$$\tag{3.13}$$

Here arrows represent the local moment orientations in a simple picture of collinear magnetism and red colour is used to help the eye. Certainly, $\tilde{\Omega}(\{\hat{e}_n\})$ is expected to be unmanageable for the evaluation of many local moment configurations in the context of SDFT calculations. Instead of working with this object, the theory presented here introduces the following trial Hamiltonian as an alternative design to describe the evolution of the local moment orientations $\{\hat{e}_n\}$,

$$\mathcal{H}_0(\{\hat{e}_n\}) = -\sum_n \mathbf{h}_n \cdot \hat{e}_n. \tag{3.14}$$

Essentially, at each magnetic site the local moment experiences the effect of a local magnetic field $\mathbf{h}_n$ that is aimed to capture the interactions described above. It can incorporate the coupling with an external magnetic field too if it is present. The Hamiltonian presented in Eq. (3.14) is our fundamental mean-field approximation and represents the physical ideas about localised exchange fields emerging from the interactions of the moving electrons and nuclei introduced earlier. Evidently, we could construct a trial Hamiltonian $\mathcal{H}_0(\{\hat{e}_n\})$ with a dependence on the local moment orientations more sophisticated than a simple single-site interaction so that the definition of local exchange field would be more complicated. Nonetheless, we shall work with Eq. (3.14) and consider the quantities $\{\mathbf{h}_n\}$ as the pertinent local magnetic fields.

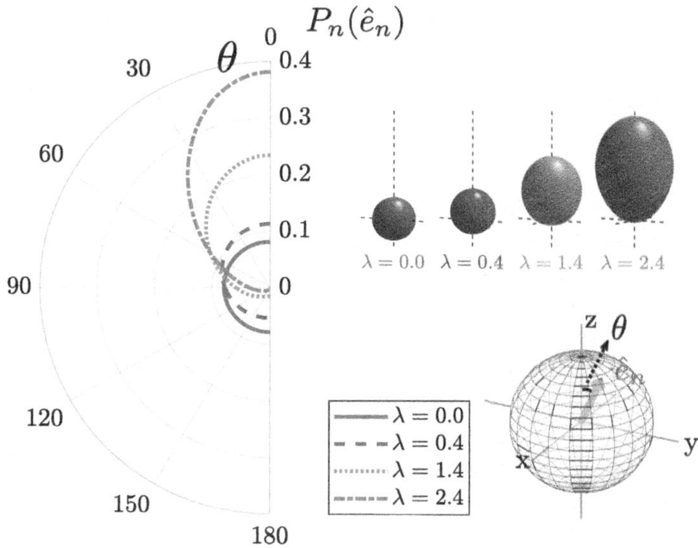

Fig. 3.1 The dependence of the single-site probability $P_n(\hat{e}_n)$ on the polar angle θ (degrees), defined with respect to the orientation of $\boldsymbol{\lambda}_n = \beta \mathbf{h}_n$ (parallel to the z-axis), for four characteristic values of $\lambda_n = |\boldsymbol{\lambda}_n|$. The figure shows that for increasing values of λ_n the shape of $P_n(\hat{e}_n)$ gradually changes from a sphere, in which all the directions in space are equally weighted, to an ellipsoid with $\hat{\boldsymbol{\lambda}}_n$ as the most preferable orientation

The average over the directions $\{\hat{e}_n\}$ can be performed consistently with $\mathcal{H}_0(\{\hat{e}_n\})$ by making use of the associated probability

$$P_0(\{\hat{e}_n\}) = \frac{1}{Z_0} \exp\left[-\beta \mathcal{H}_0(\{\hat{e}_n\})\right] = \prod_n P_n(\hat{e}_n), \tag{3.15}$$

where $P_n(\hat{e}_n)$ are the single-site probabilities

$$P_n(\hat{e}_n) = \frac{\exp\left[\beta \mathbf{h}_n \cdot \hat{e}_n\right]}{Z_{0,n}}, \tag{3.16}$$

and Z_0 is the corresponding partition function,

$$Z_0 = \prod_n Z_{0,n} = \prod_n \int d\hat{e}_n \exp\left[\beta \mathbf{h}_n \cdot \hat{e}_n\right] = \prod_n 4\pi \frac{\sinh \beta h_n}{\beta h_n}. \tag{3.17}$$

By inspecting Eqs. (3.16) and (3.17) it can be seen that the shape of the single-site probabilities depend on the direction and length of $\boldsymbol{\lambda}_n = \beta \mathbf{h}_n$ at each site n, as shown in Fig. 3.1. Note that all quantities factor due to the single-site nature of the Hamiltonian $\mathcal{H}_0$ and, therefore, the fields $\{\mathbf{h}_n\}$ are independent quantities. We can use now Eq. (3.16) to carry out the average over $\{\hat{e}_n\}$ of other objects of interest. For example, the averaged values of $\{\hat{e}_n\}$ are calculated by performing the integrals

$$\left\{ \mathbf{m}_n = \int d\hat{e}_n \, P_n(\hat{e}_n)\hat{e}_n = \left(\frac{-1}{\beta h_n} + \coth \beta h_n \right) \hat{h}_n \right\}. \tag{3.18}$$

The quantities $\{\mathbf{m}_n\}$ describe the amount of magnetic ordering at every magnetic site associated with the orientational configurations of the local moments. In other words, they are the average of the local moments orientations at long time scales in which the system has enough time to explore all possible directions via thermal fluctuations. They are the local order parameters of the single-site normalised magnetisations and we name them magnetic order parameters. Thus, a magnetic phase is fully specified by the set $\{\mathbf{m}_n\}$. For example, the paramagnetic state corresponds to $\{\mathbf{m}_n\} = \{\mathbf{0}\}$ and a ferromagnetic state to $\{\mathbf{m}_n\} = \{\mathbf{m}_{\mathrm{FM}}\}$. Moreover, a helical antiferromagnetic ordering modulated by a wave vector $\mathbf{q}_0 = (0, 0, q_0)$ applies when $\mathbf{m}_n = m_0 \left[\cos(\mathbf{q}_0 \cdot \mathbf{R}_n), \sin(\mathbf{q}_0 \cdot \mathbf{R}_n), 0 \right]$, where $\mathbf{R}_n$ is the position of the nth ferromagnetic layer perpendicular to $\mathbf{q}_0$.

The Gibbs free energy associated with $\mathcal{H}_0$ is given by

$$\mathcal{G}_0 = \frac{-1}{\beta} \log Z_0. \tag{3.19}$$

Since $\mathcal{H}_0$ is a trial Hamiltonian, according to the Peierls–Feynman inequality [10, 11] an upper bound $\mathcal{G}_1$ of the exact Gibbs free energy $\mathcal{G}$, associated with the Hamiltonian $\tilde{\Omega}(\{\hat{e}_n\})$, can be calculated as

$$\mathcal{G}_1 = \mathcal{G}_0 + \langle \tilde{\Omega}(\{\hat{e}_n\}) - \mathcal{H}_0(\{\hat{e}_n\})\rangle_0 \geq \mathcal{G}, \tag{3.20}$$

where $\langle \cdots \rangle_0$ is the ensemble average with respect to the trial probability distribution $P_0(\{\hat{e}_n\})$. To develop further Eq. (3.20) it is necessary to calculate the average of $\tilde{\Omega}(\{\hat{e}_n\})$ with respect to $P_0(\{\hat{e}_n\})$. Under the presence of an external magnetic field $\mathbf{H}$, that couples with local moments with sizes $\{\mu_n\}$ at every magnetic site n, this can be written as

$$\langle \tilde{\Omega}(\{\hat{e}_n\})\rangle_0 = \langle \Omega^{\mathrm{int}} - \sum_n \mu_n \hat{e}_n \cdot \mathbf{H}\rangle_0 = \langle \Omega^{\mathrm{int}}\rangle_0 - \sum_n \mu_n \mathbf{m}_n \cdot \mathbf{H}. \tag{3.21}$$

where we have explicitly separated the term accounting for the coupling with $\mathbf{H}$. Thus, Eq. (3.21) defines the object Ω^{int} as $\tilde{\Omega}(\{\hat{e}_n\})$ in the presence of no external magnetic field. Evidently, it depends on $\{\hat{e}_n\}$ although we have shortened the notation. After some simple algebra it can be shown from Eqs. (3.14), (3.16), and (3.19) that

$$\mathcal{G}_0 - \langle \mathcal{H}_0\rangle_0 = \frac{1}{\beta} \langle \log P(\{\hat{e}_i\})\rangle_0 = -T S_{\mathrm{mag}}, \tag{3.22}$$

where

$$S_{\text{mag}} = \sum_n S_n(\beta h_n) \tag{3.23}$$

is the magnetic entropy contribution from the orientational configurations of the local moments. Due to the single-site nature of our trial Hamiltonian S_{mag} is composed by the single-site magnetic entropies

$$\begin{aligned}
S_n(\beta h_n) &= -k_{\text{B}} \int \mathrm{d}\hat{e}_n \, P_n(\hat{e}_n) \log P_n(\hat{e}_n) \\
&= k_{\text{B}} \left[1 + \log \left(4\pi \frac{\sinh \beta h_n}{\beta h_n} \right) - \beta h_n \coth \beta h_n \right].
\end{aligned} \tag{3.24}$$

It is useful to show after some manipulations of Eq. (3.24) that

$$\mathbf{h}_n = \frac{\partial \left[-T S_{\text{mag}} \right]}{\partial \mathbf{m}_n}. \tag{3.25}$$

From Eqs. (3.20), (3.21), and (3.22) we can finally write the following expression for the upper bound of the Gibbs free energy

$$\mathcal{G}_1 = \langle \Omega^{\text{int}} \rangle_0 - \sum_n \mu_n \mathbf{m}_n \cdot \mathbf{H} - T S_{\text{mag}}. \tag{3.26}$$

Since the average over the magnetic configurations is taken with respect to $P_0(\{\hat{e}_n\})$, the natural order parameters of $\mathcal{G}_1$ are $\{\mathbf{m}_n\}$. Of course, in the equilibrium $\mathcal{G}_1$ has to be minimised with respect to them. To ensure this we take the derivative of Eq. (3.26) with respect to $\mathbf{m}_n$ and use Eq. (3.25) to write

$$- \nabla_{\mathbf{m}_n} \mathcal{G}_1 = \mathbf{h}_n^{\text{int}} + \mu_n \mathbf{H} - \mathbf{h}_n = \mathbf{0}, \tag{3.27}$$

which defines the internal magnetic fields

$$\left\{ \mathbf{h}_n^{\text{int}} = -\frac{\partial \langle \Omega^{\text{int}} \rangle_0}{\partial \mathbf{m}_n} \right\}. \tag{3.28}$$

The equilibrium condition, therefore, can also be expressed as

$$\left\{ \mathbf{h}_n = \mathbf{h}_n^{\text{int}} + \mu_n \mathbf{H} \right\}. \tag{3.29}$$

This result together with Eq. (3.18) are the central equations describing the magnetic state in equilibrium. It is interesting to examine the status of the theory at this point. We emphasise that from the perspective of Eqs. (3.14) and (3.18), $\mathbf{h}_n$ should be regarded as the exchange field that, if localised at site n, would sustain the n-th local moment with an averaged orientation equal to $\mathbf{m}_n$. Moreover, the physical meaning of $\mathbf{h}_n^{\text{int}}$ is given by Eq. (3.28), which shows that it is the emerging local magnetic field at site n when the electronic structure is forced to coexist with a magnetic ordering prescribed by $\{\mathbf{m}_n\}$, as imposed by the averaging over $P_0(\{\hat{e}_n\})$. Hence, Eq. (3.29) has a clear physical interpretation: In the equilibrium the total local magnetic field, composed by the addition of $\mathbf{h}_n^{\text{int}}$ and $\mathbf{H}$, must be identical at every site to the magnetic field necessary to sustain the local moments whose fluctuating magnetic orientations are, in average, $\{\mathbf{m}_n\}$.

Note that a given magnetic state is specified by the order parameters $\{\mathbf{m}_n\}$, or equivalently by the probabilities $\{P_n(\hat{e}_n)\}$ which in turn are prescribed by $\{\boldsymbol{\lambda}_n\} = \{\beta\mathbf{h}_n\}$ (see Eqs. (3.16) and (3.18)). The strategy is to use the KKR-DFT formalism, together with the CPA to carry out the averages over the magnetic configurations consistent with $\mathcal{H}_0$, to calculate the right hand side of Eq. (3.28) and, consequently, extract $\{\mathbf{h}_n^{\text{int}}\}$ for an input of $\{\mathbf{m}_n\}$ (or $\{\beta\mathbf{h}_n\}$) [6]. Hence, our theory is designed to describe the dependence of the local magnetic fields on the state of magnetic order. As will be explained in Chap. 4, this calculation and Eq. (3.18) are the basis of a scheme to find the most stable magnetic state as well as to construct magnetic phase diagrams against temperature and other external influences. We point out that a simple pairwise model with constants $\{J_{ij}\}$ would map into a magnetic dependence described as $\mathbf{h}_i^{\text{int}}(\{\mathbf{m}_n\}) = \sum_j J_{ij}\mathbf{m}_j$. However, multi-spin interactions are important for the magnetic materials studied in Chaps. 5 and 6 and, in consequence, the internal magnetic fields will show a non-linear behaviour.

To conclude this section we give some detail about a magnetic state that is out of the equilibrium, i.e. $\nabla_{\mathbf{m}_n}\mathcal{G}_1 \neq \mathbf{0}$ at one atomic site at least. In this situation, the values of $\{\mathbf{m}_n\}$ do not minimise $\mathcal{G}_1$ and Eq. (3.29) is not satisfied. It is instructive to express Eq. (3.27) more generally in spherical coordinates as

$$-\nabla_{\mathbf{m}_n}\mathcal{G}_1|_{\text{non-eq.}} = \frac{1}{m_n}\mathbf{T}_{\theta_n} + \frac{1}{m_n\sin\theta_n}\mathbf{T}_{\phi_n} - \frac{\partial\mathcal{G}_1}{\partial m_n}\hat{m}_n\bigg|_{\text{non-eq.}} \neq \mathbf{0} \quad (3.30)$$

$$-\nabla_{\mathbf{m}_n}\mathcal{G}_1|_{\text{eq.}} = \mathbf{0}, \quad (3.31)$$

where $\hat{m}_n = \mathbf{m}_n/m_n$ and

$$\mathbf{T}_{\theta_n} = -\frac{\partial\mathcal{G}_1}{\partial\theta_n}\hat{\theta}, \quad \mathbf{T}_{\phi_n} = -\frac{\partial\mathcal{G}_1}{\partial\phi_n}\hat{\phi} \quad (3.32)$$

are the components of the induced torque along the magnetisation angular directions $\hat{\theta}$ and $\hat{\phi}$. We stress that these components are non-zero only if the Gibbs free energy is not minimised. The physical picture is as follows. If the magnetic ordering is compatible with its electronic environment and the system is, therefore, in equilibrium ($\nabla_{\mathbf{m}_n}\mathcal{G}_1 = 0$), the local moment directions, emerging from the rapidly changing electronic structure, thermally fluctuate such that the magnetic state is stabilised at long periods of time with fixed values of $\{\mathbf{m}_n\}$. Contrary, the resulting effect when the nature of the fast evolving electronic glue cannot sustain the magnetic ordering ($\nabla_{\mathbf{m}_n}\mathcal{G}_1 \neq 0$) is that the total magnetic field at each site, $\mathbf{h}_n^{\text{int}} + \mu_n\mathbf{H}$, might be different to the local field $\mathbf{h}_n$ necessary to sustain the magnetic ordering $\{\mathbf{m}_n\}$. As a consequence, the local moments experience the effect of out-of-equilibrium magnetic fields given by Eq. (3.30) and, therefore, the magnetic ordering is unstable to the formation of another more preferable magnetic state.

3.3 The Internal Local Field and Its Electronic Glue Origins

The theory presented in Sect. 3.2 grounds the route to obtain an upper-bound of the Gibbs free energy from the statistical mechanics of disordered local moments within a mean field approximation. However, the internal magnetic energy $\langle \Omega^{\text{int}} \rangle_0$ is yet unknown. The evaluation of this object constitutes the electronic structure part of the theory and ultimately shall lead to the calculation of the local internal fields, $\{\mathbf{h}_n^{\text{int}}\}$, which we explain how to calculate in this section.

3.3.1 The Coherent Potential Approximation of Disordered Local Moments

Carrying out the ensemble averages over $\{\hat{e}_n\}$ is yet the task that shows itself as intractable. In principle, the way to proceed in order to evaluate exactly the grand potential Ω would require to calculate $\tilde{\Omega}(\{\hat{e}_n\})$ for many magnetic configurations and then use Eqs. (3.10) and (3.11) to perform the average with respect to $P(\{\hat{e}_n\})$. Fortunately, the mean field theory presented in Sect. 3.2 establishes the means to approximate this calculation yielding a scheme that is computationally affordable. The idea is to exploit the single-site nature of the equilibrium probability associated with the trial Hamiltonian, $P_0(\{\hat{e}_n\})$, and circumvent the calculation of Ω by alternatively evaluating $\langle \tilde{\Omega}(\{\hat{e}_n\}) \rangle_0$. Of course, this procedure is realised within the framework of the mean field approach presented in Sect. 3.2 and, therefore, it is suitable for the calculation of the upper bound of the Gibbs free energy $\mathcal{G}_1$.

To illustrate this strategy, we focus our attention on the self-consistent cycle of a SDFT calculation for the problem set by $\tilde{\Omega}(\{\hat{e}_n\})$ with a magnetic configuration $\{\hat{e}_n\}$. At every cycle the potentials are supposed to be reconstructed from the electron and local moment densities until a self-consistent solution is obtained. The complexity lies in that the densities are functions of the entire set of magnetic directions $\{\hat{e}_n\}$ so that the potentials depend on them too. Our way to proceed is to implement the CPA, presented in Sect. 2.3 as the technique to describe compositional disorder in alloys, to carry out the average over $\{\hat{e}_n\}$ and, hence, simplify the construction of the potentials at each iteration. The idea is that our mean field approach naturally sets the single-site probabilities that will be used in the CPA method. The approximation consists of replacing the disordered potential at each site by an effective medium that approximates as best as possible the effect of the averaged random potential of disordered local moments. Analogously to the atomic disorder case, the prescription is that the average of the scattering effects from the thermal fluctuations of a local moment when is embedded by the effective medium is identical to those ones caused by the effective medium itself.

We now proceed to derive the CPA equations. By recalling Eq. (2.81), the CPA condition, consistent with $P_0(\{\hat{e}_n\})$, can be expressed in the language of MST as

$$\left\{ \int \mathrm{d}\hat{e}_n \, P_n(\hat{e}_n)\underline{t}_n^{\mathrm{diff}}(\hat{e}_n) = 0 \right\}, \tag{3.33}$$

where $\underline{t}_n^{\mathrm{diff}}(\hat{e}_n)$ describes the effect of replacing the effective medium by a local moment oriented along $\hat{e}_n$ at site n. Note that due to the continuous range of possible values of the magnetic orientation the averages are now performed as integrals. Following a similar reasoning as presented in Sect. 2.3, from Eq. (2.81) one can obtain an expression for the total scattering T-matrix of the effective medium, $\underline{T}_c$, in terms of the T-matrix associated with a system where the magnetic sites contain disordered local moments pointing along $\{\hat{e}_n\}$, $\underline{T}(\{\hat{e}_n\})$,

$$\underline{T}_c = \prod_n \int \mathrm{d}\hat{e}_n \, P_n(\hat{e}_n)\underline{T}(\{\hat{e}_n\}). \tag{3.34}$$

From this equation it follows that the diagonal parts of the corresponding SPOs for an arbitrary site n_0 are related as

$$\underline{\tau}_{c,n_0n_0} = \prod_n \int \mathrm{d}\hat{e}_n \, P_n(\hat{e}_n)\underline{\tau}_{n_0n_0}(\{\hat{e}_n\}) = \int \mathrm{d}\hat{e}_{n_0} P_{n_0}(\hat{e}_{n_0})\langle\underline{\tau}_{n_0n_0}\rangle_{\hat{e}_{n_0}}, \tag{3.35}$$

where $\langle\cdots\rangle_{\hat{e}_{n_0}}$ indicates that the average is performed at every site apart from n_0. We recall that $\underline{\tau}_{n_0n_0}(\{\hat{e}_n\})$ and $\underline{\tau}_{c,n_0n_0}$ are constructed from the free-particle Green's function, containing the structural information, and the appropriate t-matrices describing the scattering of disordered local moments, $\{\underline{t}_n(\hat{e}_n)\}$, and the coherent potential, $\{\underline{t}_{c,n}\}$, respectively. Equation (2.66) can be invoked to find from Eq. (3.35) the fundamental CPA equations of disordered local moments. We firstly write

$$\underline{\tau}_{c,n_0n_0}^{-1} = \underline{\tau}_{n_0n_0}^{-1}(\{\hat{e}_n\}) - \underline{t}_{n_0}^{-1}(\hat{e}_{n_0}) + \underline{t}_{c,n_0}^{-1}, \tag{3.36}$$

which after simple manipulations results in

$$\langle\underline{\tau}_{n_0n_0}\rangle_{\hat{e}_{n_0}} = \underline{D}_{n_0}(\hat{e}_{n_0})\underline{\tau}_{c,n_0n_0}, \tag{3.37}$$

where the impurity matrix is

$$\underline{D}_{n_0}(\hat{e}_{n_0}) = \left[\underline{1} + \left(\underline{t}_{n_0}^{-1}(\hat{e}_{n_0}) - \underline{t}_{c,n_0}^{-1}\right)\underline{\tau}_{c,n_0n_0}\right]^{-1} \tag{3.38}$$

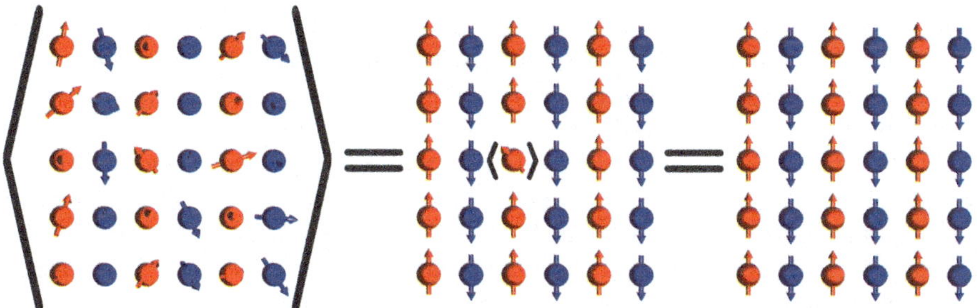

Fig. 3.2 The CPA of a two-dimensional system with two non-equivalent magnetic positions (red and blue) whose fluctuating magnetic directions are oriented oppositely after the average is performed. The figure shows that when the red moment, for example, is surrounded by the effective medium, which is represented by the average orientations, its fluctuation produces in average the same scattering effects as the effective medium itself. The same condition has to be fulfilled by the blue local moment so that the resulting effective medium is constructed from the CPA prescription at every magnetic site self-consistently

It is a matter of simple algebra to show from the CPA condition in Eq. (3.35) that the CPA equations for disordered local moments are

$$\left\{ \int d\hat{e}_n \, P_n(\hat{e}_n) \underline{D}_n(\hat{e}_n) = \underline{1} \right\}, \tag{3.39}$$

or the numerically more convenient version

$$\left\{ \int d\hat{e}_n \, P_n(\hat{e}_n) \underline{X}_n(\hat{e}_n) = \underline{0} \right\}, \tag{3.40}$$

with the excess scattering matrix now being

$$\underline{X}_n(\hat{e}_n) = \left[\left(\underline{t}_n^{-1}(\hat{e}_n) - \underline{t}_{c,n}^{-1} \right)^{-1} + \underline{\tau}_{c,nn} \right]^{-1}. \tag{3.41}$$

The implementation of the CPA equations is, therefore, self-consistent with each magnetic site for magnetic unit cells containing an arbitrary number of non-equivalent magnetic sites. To illustrate this, we show in Fig. 3.2 the CPA of disordered local moments for a magnetic structure with two non-equivalent magnetic sites whose averaged magnetic directions align anti parallel.

3.3.2 Self-consistent Calculations

Here we describe the self-consistent iterative procedure to solve the Kohn-Sham equations within the mean field approach for disordered local moments and give some detail on its numerical implementation. The central objects used at the beginning of every iteration are, of course, the electron and magnetic moment densities, $n(\mathbf{r})$ and $\mu(\mathbf{r})$. Due to the muffin-tin form of the potentials, it is convenient to associate single-site densities $n_n(\mathbf{r})$ and $\mu_n(\mathbf{r})$ with each site n, directly given by restricting the spatial coordinate to the surrounding volume. Our KKR-MST formalism is designed to obtain them by using Eqs. (2.69) and (2.70), respectively, which requires the calculation of the Green's function given by the SPO and Eq. (2.67). Following the CPA ideology, we consider the partially averaged densities $\langle n_n(\mathbf{r})\rangle_{\hat{e}_n}$ and $\langle \mu_n(\mathbf{r})\rangle_{\hat{e}_n}$ at each site in line with the mean field medium to be constructed at every iteration. They are obtained by using the same equations but with a Green's function calculated from the partial average of the SPO $\langle \underline{\tau}_{n_0 n_0}\rangle_{\hat{e}_n}$, given in Eq. (3.37), that is

$$\langle n_n(\mathbf{r})\rangle_{\hat{e}_n} = \mp\frac{1}{\pi}\mathrm{ImTr}\int dE f(E)\langle G^{\pm}(\mathbf{r}, \mathbf{r}, E)\rangle_{\hat{e}_n}, \tag{3.42}$$

$$\langle \mu_n(\mathbf{r})\rangle_{\hat{e}_n} = \pm\mu_{\mathrm{B}}\frac{1}{\pi}\mathrm{ImTr}\int dE f(E)\beta_I \mathbf{\Sigma}\langle G^{\pm}(\mathbf{r}, \mathbf{r}, E)\rangle_{\hat{e}_n}, \tag{3.43}$$

where

$$\langle G(\mathbf{r}, \mathbf{r}, E)\rangle_{\hat{e}_n} = \underline{Z}_n(\mathbf{r} - \mathbf{R}_n, E)\langle \underline{\tau}_{nn}(E)\rangle_{\hat{e}_n}\underline{Z}_n^{\times}(\mathbf{r} - \mathbf{R}_n, E) \\ -\underline{Z}_n(\mathbf{r} - \mathbf{R}_n, E)\underline{J}_n^{\times}(\mathbf{r} - \mathbf{R}_n, E). \tag{3.44}$$

Now all is set to define the self-consistent calculation of a disordered local moment state at finite temperature. For a chosen set of single-site probabilities $\{P_n(\hat{e}_n)\}$ and a first guess of the partially averaged densities, we proceed by performing the following steps

(i) We firstly construct the single-site muffin-tin potentials at every site. By recalling the definitions inside the effective SDFT potential in Chap. 2 (see for example Eqs. (2.13), (2.42), and (2.41)), within the mean field approach we write

$$\langle V_n(\mathbf{r}_n)\rangle_{\hat{e}_n} = V_{\mathrm{ext}}(r_n) + \left(V_{\mathrm{xc}}(\mathbf{r}) - \mu_{\mathrm{B}}\sigma\cdot\mathbf{B}_{\mathrm{xc}}(\mathbf{r})\right)\Big|_{\langle n_n(\mathbf{r})\rangle_{\hat{e}_n}, \langle\mu_n(\mathbf{r})\rangle_{\hat{e}_n}}$$
$$+ e^2\int_{V_n} d\mathbf{r}'\,\frac{\langle n_n(\mathbf{r})\rangle_{\hat{e}_n}}{4\pi\epsilon_0|\mathbf{r} - \mathbf{r}'|} + e^2\sum_{m\neq n}\int_{V_m} d\mathbf{r}'\,\frac{\langle n_m(\mathbf{r})\rangle}{4\pi\epsilon_0|\mathbf{r} - \mathbf{R}_n - \mathbf{r}' + \mathbf{R}_m|}, \tag{3.45}$$

where the external potential, containing the effect from the fixed nuclei, follows from the muffin-tin construction, the second and third terms are given by Eq. (2.42) evaluated with $\langle n_n(\mathbf{r})\rangle_{\hat{e}_n}$ and $\langle \mu_n(\mathbf{r})\rangle_{\hat{e}_n}$ and the LDA, and the last two terms form the Hartree potential. Note that the fully averaged electron density is directly calculated as

$$\langle n_n(\mathbf{r})\rangle = \int d\hat{e}_n\, P_n(\hat{e}_n)\langle n_n(\mathbf{r})\rangle_{\hat{e}_n}. \tag{3.46}$$

(ii) Solve the single-site scattering problem from the potentials obtained in (i) at each site to find the t-matrices $\{\underline{t}_n(\hat{e}_n)\}$ and spatial solutions $\{\underline{Z}_n(\mathbf{r}, E)\}$ and $\{\underline{J}_n(\mathbf{r}, E)\}$. We point out that owing to the muffin-tin shape of the potentials, the spherical symmetry can be exploited such that the dependence on the magnetic orientation of $\{\underline{t}_n(\hat{e}_n)\}$ can be simply extracted with

$$\underline{t}_n(\hat{e}_n) = \underline{R}(\hat{e}_n)\underline{t}_n^{\text{ref}}\underline{R}(\hat{e}_n)^+, \tag{3.47}$$

where $\underline{t}_n^{\text{ref}}$ is the t-matrix expressed in a local spin frame of reference in which it is diagonal, i.e. in which the $\hat{z}$-axis is parallel to $\hat{e}_n$, and $\underline{R}(\hat{e}_n)$ is a unitary representation of the $O(3)$ transformation that rotates the frame of reference at site n along $\hat{e}_n$,[2]

$$\underline{t}_n^{\text{ref}} = \begin{pmatrix} \underline{t}_{n+} & \underline{0} \\ \underline{0} & \underline{t}_{n-} \end{pmatrix} \Rightarrow \underline{t}_n(\hat{e}_n) = \frac{1}{2}\left(\underline{t}_{n+} + \underline{t}_{n-}\right)\underline{I}_2 + \frac{1}{2}\left(\underline{t}_{n+} - \underline{t}_{n-}\right)\boldsymbol{\sigma}\cdot\hat{e}_n. \tag{3.48}$$

(iii) Use $\{\underline{t}_n(\hat{e}_n)\}$ and $\{P_n(\hat{e}_n)\}$, as well as the structural information, to solve the CPA equations given by Eq. (3.40) and, hence, obtain $\{\underline{t}_{c,n}\}$ and $\{\underline{\tau}_{c,nn}\}$ defining the effective medium.

(iv) Use Eqs. (3.42) and (3.43), together with Eqs. (3.37) and (3.44), to recalculate the single-site partially averaged densities.

(v) Check if $\langle n_n(\mathbf{r})\rangle_{\hat{e}_n}$ and $\langle \mu_n(\mathbf{r})\rangle_{\hat{e}_n}$ are the same as the initial guesses within a satisfactory numerical accuracy. If not, start again from (i) using the result from (iv).

In principle, the self-consistent procedure described above is meant to be carried out for each set of probabilities $\{P_n(\hat{e}_n)\}$ of interest. Nonetheless, if the size of the magnetic moment at every site has little dependence on the state of magnetic order $\{\mathbf{m}_n\}$ then the computation can be simplified by assuming that $\{\underline{t}_n^{\text{ref}}\}$ have no dependence on $\{P_n(\hat{e}_n)\}$ so that the calculation of $\{\underline{t}_n(\hat{e}_n)\}$ for a single magnetic configuration is sufficient. Once they are calculated there is only need to perform steps (iii)–(iv) for different magnetic orderings prescribed by $\{P_n(\hat{e}_n)\}$. This is the known Rigid Spin Approximation (RSA) and when it applies it is usually referred to the presence of good local moments, which in principle arise from fairly well localised d and f electrons. Evidently, the validity of this approximation can be tested by proceeding to the next iteration and hence checking the self-consistency of potentials and electron and magnetic moment densities.

Once self-consistent potentials and densities are obtained by employing such a scheme, we can proceed to the evaluation of $\langle \Omega^{\text{int}}\rangle_0$. A formal way to derive this object follows by making use of the chemical potential for a constrained magnetic configuration $\{\hat{e}_n\}$ to calculate

---

[2] Note that $\underline{R} = \underline{I}_2 \cos(\theta_{\hat{e}_n}/2) - i\sigma_j \sin(\theta_{\hat{e}_n}/2)$, where $\theta_{\hat{e}_n}$ is the angle between $\hat{e}_n$ and the $\hat{z}$-axis, in the frame of reference of $\underline{t}_n^{\text{ref}}$, and σ_j indicates the direction of rotation.

$$\Omega^{\mathrm{int}}(\{\hat{e}_n\}) = -\int_{-\infty}^{\nu} \mathrm{d}\nu' N(\nu'; \{\hat{e}_n\}), \tag{3.49}$$

and then perform the averaging over all magnetic configurations,

$$\langle \Omega^{\mathrm{int}} \rangle_0 = \prod_n \int \mathrm{d}\hat{e}_n \, P_n(\hat{e}_n) \Omega^{\mathrm{int}}(\{\hat{e}_n\}) = -\prod_n \int \mathrm{d}\hat{e}_n \, P_n(\hat{e}_n) \int_{-\infty}^{\nu} \mathrm{d}\nu' N(\nu'; \{\hat{e}_n\}). \tag{3.50}$$

After some simple algebra and by using the definitions given in Eqs. (2.72) and (2.73), we can write Eq. (3.50) as

$$\langle \Omega^{\mathrm{int}} \rangle_0 = \int_{-\infty}^{\infty} \mathrm{d}E \, \bar{N}(E) f(E-\nu) - \int_{-\infty}^{\infty} \mathrm{d}E \int_{-\infty}^{\nu} \mathrm{d}\nu' \frac{\partial \bar{N}(E)}{\partial \nu'} f(E-\nu) \tag{3.51}$$

where

$$\bar{N}(E) \equiv \prod_n \int \mathrm{d}\hat{e}_n \, P_n(\hat{e}_n) \int \mathrm{d}E \, n(E). \tag{3.52}$$

The last term in the right-hand side of Eq. (3.51) can be shown to be negligible compared with the first term [6]. At this point it is only necessary to take the derivative of this result with respect to the magnetic order parameters to calculate the internal magnetic fields (see Eq. (3.28)) [6, 12],

$$\mathbf{h}_n^{\mathrm{int}} = -\frac{1}{\pi}\mathrm{Im} \int \mathrm{d}\hat{e}_n \, \frac{\partial P_n(\hat{e}_n)}{\partial \mathbf{m}_n} \int_{-\infty}^{\infty} \mathrm{d}E f(E-\nu) \log \det \underline{D}_n(\hat{e}_n), \tag{3.53}$$

where Eq. (3.37) as well as the Lloyd formula expressed with [13]

$$\delta N(E) = \pm \frac{1}{\pi} \mathrm{Im} \log \det \underline{\tau}(E) \tag{3.54}$$

have been used.

3.4 Wave-Modulated Magnetic Structures from Fully Disordered Local Moments

We can investigate the stability of different magnetic phases at finite temperatures by calculating the corresponding internal local fields emerging from the electronic structure as a response to the magnetic ordering itself. However, the magnetic interactions might favour very complicated magnetic arrangements, such as non-collinear phases and/or wave-modulated incommensurate magnetism described by several crystallographic cells, for example. Studying all the possible competing magnetic ordering can be a formidable mathematical challenge very difficult to solve from the calculation of $\{\mathbf{h}_n^{\text{int}}\}$ only. For example, it would demand to construct large magnetic unit cells and examine the magnetic interactions in real space. To overcome this problem it is useful to define the so-called direct correlation function, which is obtained by taking the second derivative of $\langle \Omega^{\text{int}} \rangle_0$ with respect to the local order parameters

$$S_{ij}^{(2)}(\{\mathbf{m}_n\}) = -\frac{\partial^2 \langle \Omega^{\text{int}} \rangle_0(\{\mathbf{m}_n\})}{\partial \mathbf{m}_i \partial \mathbf{m}_j} = \frac{\partial \mathbf{h}_i^{\text{int}}(\{\mathbf{m}_n\})}{\partial \mathbf{m}_j}. \tag{3.55}$$

The strategy is to investigate the instabilities of the fully disordered high temperature paramagnetic state in the reciprocal space and gain information of the type of magnetic ordering that might occur when small magnetic polarisations $\{\delta\mathbf{m}_n\}$ are induced at each site n. We accomplish this by studying the effect of applying an infinitesimally small site-dependent external magnetic field $\mathbf{H}_n$,

$$\delta\mathbf{m}_i = \sum_j \chi_{ij}(\{\mathbf{m}_n\})\mathbf{H}_j, \tag{3.56}$$

where the linear response to the magnetic field application is described by the magnetic susceptibility $\chi_{ij}(\{\mathbf{m}_n\})$. By using Eq. (3.55) and recalling that $\mathbf{h}_n = \mathbf{h}_n^{\text{int}} + \mathbf{H}_n$ in the equilibrium (see Eq. 3.29) we can write

$$\frac{\partial \mathbf{m}_i}{\partial \mathbf{H}_j} = \frac{\partial \mathbf{m}_i}{\partial \mathbf{h}_i}\left(\delta_{ij} + \sum_k S_{ik}^{(2)}\frac{\partial \mathbf{m}_k}{\partial \mathbf{H}_j}\right). \tag{3.57}$$

Expressing this equation in terms of the magnetic susceptibility gives

$$\chi_{ij} = \chi_{0,i}\delta_{ij} + \chi_{0,i}\sum_k S_{ik}^{(2)}\chi_{kj}, \tag{3.58}$$

which defines the molecular magnetic susceptibility $\chi_{0,n}$ as the change of the local order parameter at site n with respect to $\mathbf{h}_n$. In our theory the paramagnetic state is composed by local moments whose orientations are fully disordered. This state corresponds to the probability distribution at each magnetic site being equally weighted at every possible orientation, that is $\{\mathbf{m}_n = \mathbf{0}\}$, or equivalently $\{\beta\mathbf{h}_n = \mathbf{0}\}$ (see Fig. 3.1). The single-site probability introduced in Eq. (3.16) then becomes

$$P_n(\hat{e}_n) \approx P_i^0 + \delta P_i(\hat{e}_l) - \frac{1}{4\pi} + \frac{3}{4\pi}\mathbf{m}_n \cdot \hat{e}_n \rightarrow \frac{1}{4\pi}. \tag{3.59}$$

In this limit it can be shown from Eqs. (3.18) and (3.59) that

$$\left\{\mathbf{m}_n \approx \frac{\beta}{3}\mathbf{h}_n\right\} \Rightarrow \chi_{0,n} \approx \frac{\beta}{3}. \tag{3.60}$$

By multiplying both sides of Eq. (3.58) by the inverted susceptibilities χ_0^{-1} and χ^{-1}, and using Eq. (3.60), it can be shown that

$$\chi_{ij} = \frac{1}{3k_{\mathrm{B}}T\delta_{ij} - S_{ij}^{(2)}},\tag{3.61}$$

where

$$\zeta_{ij} = \frac{\partial^2 \mathcal{G}_1}{\partial \mathbf{m}_i \partial \mathbf{m}_j} = 3k_{\mathrm{B}}T\delta_{ij} - S_{ij}^{(2)}\tag{3.62}$$

are the components of the Hessian matrix ζ_{ij} associated with the Gibbs free energy $\mathcal{G}_1$ introduced in Eq. (3.26). The highest transition temperature $T_{\max}$ at which the paramagnetic state is unstable to the formation of a magnetic phase can be calculated by solving the condition for the magnetic susceptibility to diverge, or equally for the determinant of ζ be zero,

$$\det\left(\zeta\right) = \det\left(\chi^{-1}\right) = 0,\tag{3.63}$$

which reduces to calculate the eigenvalues $\{u_a\}$ of $S_{ij}^{(2)}$ satisfying

$$\prod_{a=1}^{N_m} [3k_{\mathrm{B}}T - u_a] = 0,\tag{3.64}$$

where N_m is the number of magnetic atoms in the crystal. Thus, $T_{\max}$ can be calculated by finding the largest eigenvalue $u_{\max}$,

$$T_{\max} = \frac{u_{\max}}{3k_{\mathrm{B}}}.\tag{3.65}$$

We can also obtain the relative orientations between the local order parameters below $T_{\max}$ by examining the equilibrium state condition $\{\nabla_{\mathbf{m}_i}\mathcal{G}_1 = \mathbf{0}\}$. When the external magnetic field is zero this condition becomes (see Eqs. (3.27) and (3.28))

$$\delta\mathbf{h}_i = -\frac{\partial\langle\Omega^{\mathrm{int}}\rangle_0}{\partial\mathbf{m}_i}\bigg|_{\{\delta\mathbf{m}_n\}}.\tag{3.66}$$

Using Eqs. (3.55) and (3.60) we can write this as

$$3k_{\mathrm{B}}T\delta\mathbf{m}_i = \sum_j S_{ij}^{(2)}\delta\mathbf{m}_j.\tag{3.67}$$

This result shows that the components of the eigenvectors of $S_{ij}^{(2)}$ are the projections of $\{\delta\mathbf{m}_n\}$ minimising $\mathcal{G}_1$ along a given direction in space.

Solving the eigenvalue problem set by $S_{ij}^{(2)}$ in the real space is in principle very complicated if there are long range magnetic interactions, requiring the diagonalisation of a high dimensional matrix as well as the calculation of its components. We follow an alternative route based on the exploitation of the translation invariance in the paramagnetic state. In order to explore the instabilities to the formation of specific magnetic structures modulated by a wave vector $\mathbf{q}$, we define the following lattice Fourier transforms

$$\tilde{S}_{ss'}^{(2)}(\mathbf{q}) = \frac{1}{N_\mathrm{c}} \sum_{tt'} S_{ts\,t's'}^{(2)} e^{i\mathbf{q}\cdot(\mathbf{R}_t+\mathbf{r}_s-\mathbf{R}_{t'}-\mathbf{r}_{s'})}, \tag{3.68}$$

$$\tilde{\chi}_{ss'}(\mathbf{q}) = \frac{1}{N_\mathrm{c}} \sum_{tt'} \chi_{ts\,t's'} e^{i\mathbf{q}\cdot(\mathbf{R}_t+\mathbf{r}_s-\mathbf{R}_{t'}-\mathbf{r}_{s'})}, \tag{3.69}$$

where N_c is the number of unit cells. We have decomposed each lattice site index i and j by two additional indices, $i \to t, s$ and $j \to t', s'$. We use t and t' to specify the origin of unit cells, and s and s' for indices denoting the atomic positions within the unit cells t and t', respectively. $\mathbf{R}_t + \mathbf{r}_s$ is, therefore, the position vector of the magnetic atom at site (t, s) such that $\mathbf{R}_t$ denotes the origin of the unit cell t and $\mathbf{r}_s$ gives the relative position of the sub-lattice s within that unit cell. Hence, $\tilde{S}_{ss'}^{(2)}(\mathbf{q})$ and $\tilde{\chi}_{ss'}(\mathbf{q})$ are square matrices whose components depend on the wave vector $\mathbf{q}$. Their dimension is the number of magnetic positions, or sub-lattices, N_at inside the unit cell, $N_\mathrm{at} = N_m/N_c$. Note that the shape and components of these matrices depend on the choice of the unit cell. Lattice Fourier transforming Eq. (3.61) gives

$$\tilde{\chi}_{ss'}(\mathbf{q}) = \frac{1}{3k_\mathrm{B}T\,\delta_{ss'} - \tilde{S}_{ss'}^{(2)}(\mathbf{q})}. \tag{3.70}$$

The condition for the magnetic susceptibility to diverge now is satisfied when

$$\det\left(\tilde{\chi}_{ss'}^{-1}\right) = 0. \tag{3.71}$$

Hence, by applying the lattice Fourier transform we have reduced the dimension of the matrix to be diagonalised from $N_m \times N_m$ to $N_\mathrm{at} \times N_\mathrm{at}$. Consequently, Eq. (3.64) becomes

$$\prod_{a=1}^{N_\mathrm{at}} \left[3k_\mathrm{B}T - \tilde{u}_a(\mathbf{q})\right] = 0, \tag{3.72}$$

where now $\{\tilde{u}_a(\mathbf{q}),\ a = 1, \ldots, N_\mathrm{at}\}$ are the N_at eigenfunctions of $\tilde{S}_{ss'}^{(2)}(\mathbf{q})$, which depend on the wave vector $\mathbf{q}$ and are obtained by solving the eigenvalue problem

$$\sum_{s'=1}^{N_\mathrm{at}} \tilde{S}_{ss'}^{(2)}(\mathbf{q}) V_{a,s'}(\mathbf{q}) = \tilde{u}_a(\mathbf{q}) V_{a,s}(\mathbf{q}), \tag{3.73}$$

for $\{s = 1, \ldots, N_{\mathrm{at}}\}$, and N_{at} eigenvectors of dimension N_{at}, $\{(V_{a,1}, \ldots, V_{a,N_{\mathrm{at}}}),$ $a = 1, \ldots, N_{\mathrm{at}}\}$. Using Eq. (3.68) we can write $\mathcal{G}_1$ without external magnetic field as

$$\mathcal{G}_1 = -N_c \sum_{\mathbf{q},ss'} \left(\tilde{S}^{(2)}_{ss'}(\mathbf{q}) - 3k_B T \delta_{ss'} \right) \delta\tilde{\mathbf{m}}_s^{a,\mathbf{q}}(\mathbf{q}) \cdot \delta\tilde{\mathbf{m}}_{s'}^{a,\mathbf{q}}(-\mathbf{q}), \tag{3.74}$$

which defines the lattice Fourier transform of the local order parameters,

$$\tilde{\mathbf{m}}_s(\mathbf{q}) = \frac{1}{N_c} \sum_t \mathbf{m}_{ts} e^{i\mathbf{q}\cdot(\mathbf{R}_t + \mathbf{r}_s)}. \tag{3.75}$$

In Eq. (3.74) the superscript $(a, \mathbf{q})$ means that when Eq. (3.75) is applied to find $\tilde{\mathbf{m}}_s(\mathbf{q})$ the quantities $\{\mathbf{m}_{ts}\}$ are modulated by $\mathbf{q}$ with orientations specified by a. The Gibbs free energy in Eq. (3.74) is minimised by the value of the wave vector $\mathbf{q}_p$ at which $\tilde{S}^{(2)}_{ss'}(\mathbf{q})$ peaks. Evidently, this is the value at which the highest value of the eigenfunctions $\{\tilde{u}_a(\mathbf{q})\}$ is found. Hence, minimising the free energy when it is expanded around the paramagnetic state yields a magnetic structure $\{\mathbf{m}_{ts}\}$ modulated by $\mathbf{q}_p$ with the highest transition temperature consistent with Eq. (3.72),

$$\max\left(\{T^{a,\mathbf{q}_a}\}\right) = T^{\mathrm{p},\mathbf{q}_p} = \frac{\tilde{u}_p(\mathbf{q}_p)}{3k_B}. \tag{3.76}$$

It is interesting to inspect the theory at this point. If the temperature is below but close to $T^{\mathrm{p},\mathbf{q}_p}$ then the system develops some finite and very small magnetic ordering prescribed by the values $\mathrm{a} = \mathrm{p}$ and $\mathbf{q}_p$,

$$\left\{ \delta\mathbf{m}_{ts}^{\mathrm{p},\mathbf{q}_p} = \delta m_{x,ts}^{\mathrm{p},\mathbf{q}_p} \hat{x} + \delta m_{y,ts}^{\mathrm{p},\mathbf{q}_p} \hat{y} + \delta m_{z,ts}^{\mathrm{p},\mathbf{q}_p} \hat{z} \right\} \tag{3.77}$$

such as the orientations of the local order parameters in the resulting magnetic state are modulated by the wave vector $\mathbf{q}_p$. In the most general form, the components along the spatial directions can be written as

$$\delta m_{x,t's}^{\mathrm{p},\mathbf{q}_p} = \delta m_{x,ts}^{\mathrm{p},\mathbf{q}_p} \cos\left(\mathbf{q}_p \cdot (\mathbf{R}_t - \mathbf{R}_{t'}) + \varphi_{x,s}^{\mathrm{p},\mathbf{q}_p}\right),$$

$$\delta m_{y,t's}^{\mathrm{p},\mathbf{q}_p} = \delta m_{y,ts}^{\mathrm{p},\mathbf{q}_p} \cos\left(\mathbf{q}_p \cdot (\mathbf{R}_t - \mathbf{R}_{t'}) + \varphi_{y,s}^{\mathrm{p},\mathbf{q}_p}\right),$$

$$\delta m_{z,t's}^{\mathrm{p},\mathbf{q}_p} = \delta m_{z,ts}^{\mathrm{p},\mathbf{q}_p} \cos\left(\mathbf{q}_p \cdot (\mathbf{R}_t - \mathbf{R}_{t'}) + \varphi_{z,s}^{\mathrm{p},\mathbf{q}_p}\right). \tag{3.78}$$

Each component of every magnetic order parameter inside the unit cell t, therefore, change by a factor $\cos\left(\mathbf{q}_p \cdot (\mathbf{R}_t - \mathbf{R}_{t'}) + \varphi_{(\hat{x},\hat{y},\hat{z}),s}^{\mathrm{p},\mathbf{q}_p}\right)$ when transferred to the unit cell t'. The parameter $\mathrm{a} = \mathrm{p}$ specifies the directions of the magnetic order parameters within one unit cell. To elucidate what these directions are we can lattice Fourier transform Eq. (3.67) along the wave vector $\mathbf{q}_p$ with magnetic order parameters modulated as in Eq. (3.78) to obtain

$$3k_{\rm B}T\,\delta\tilde{\mathbf{m}}_s^{\rm p,\mathbf{q}_{\rm p}}(\mathbf{q}_{\rm p}) = \sum_{s'=1}^{N_{\rm at}} \tilde{S}_{ss'}^{(2)}(\mathbf{q}_{\rm p})\delta\tilde{\mathbf{m}}_{s'}^{\rm p,\mathbf{q}_{\rm p}}(\mathbf{q}_{\rm p}),\tag{3.79}$$

Similarly to Eqs. (3.67) and (3.79) shows that the components of the eigenvectors of $\tilde{S}_{ss'}^{(2)}$ have to be interpreted as the projections of the Lattice Fourier transformed local order parameters $\{\delta\tilde{\mathbf{m}}_s(\mathbf{q})\}$ along a given direction in space $\hat{r}$,

$$(V_{a,1}, \ldots, V_{a,N_{\rm at}}) = (\tilde{m}_{\hat{r},1}, \ldots, \tilde{m}_{\hat{r},N_{\rm at}}).\tag{3.80}$$

Hence, the eigenvectors, prescribed by the number a, contain the information of the relative orientations between the $N_{\rm at}$ magnetic order parameters inside one unit cell t, $\{\delta\mathbf{m}_{t1}, \ldots, \delta\mathbf{m}_{tN_{\rm at}}\}$, which in turn rotate to another unit cell t' by the modulation mediated by the wave vector $\mathbf{q}_{\rm p}$, as described in Eq. (3.78). For example, when $\mathbf{q}_{\rm p} = \mathbf{0}$ the paramagnetic state is unstable to the formation of ferromagnetic sub-lattices, i.e., the magnetic order parameters within each sub-lattice are parallel and they do not rotate from one unit cell to another. Note that the magnetic polarisations between different sub-lattices are aligned only if all the components of the eigenvector are equal, as shown in Fig. 3.3a. On the other hand, if the eigenfunction peaks at the Brillouin zone along the direction $\mathbf{q}_{\rm p} = (q_{\rm BZ}, 0, 0)$ then the local order parameters completely reverse their orientations when changing from one unit cell to the next one along the $\hat{x}$ direction while they maintain the same orientation when transferred along the $\hat{y}$ and $\hat{z}$ directions. In Fig. 3.3 we show this situation in panel (d) as well as other cases in panels (b) and (c) for illustrative purposes. As a final remark, we point out that longer periods occur when $\mathbf{q}_{\rm p}$ lies inside the Brillouin zone.

The way to proceed from this analysis is as follows. We guide the internal field calculation of stable structures described in the previous section by firstly inspecting at which values of $\mathbf{q}$ the eigenfunctions of $\tilde{S}_{ss'}^{(2)}(\mathbf{q})$ peak. The idea is to examine the components of the corresponding eigenvectors to obtain the relative orientations between the local order parameters at different magnetic sub-lattices. This allows for the extraction of the magnetic structures that compete at high temperatures.

We identify them as the magnetic structures of interest and investigate the effect of lowering the temperature on their stabilisation by the calculation of the internal local fields $\{\mathbf{h}_n^{\rm int}\}$ at different values of the magnetic ordering $\{\mathbf{m}_n\}$.

3.4.1 Calculation of the Direct Correlation Function

We conclude this chapter by showing how to obtain $\tilde{S}_{ss'}^{(2)}(\mathbf{q})$ from our DFT-DLM formalism. The way to proceed follows by recalling $\mathbf{h}_n^{\rm int}$ expressed in terms of the impurity matrix, given in Eq. (3.53). By taking the derivative with respect to the local order parameter and by using Eqs. (3.55) and (3.59) we obtain the direct correlation

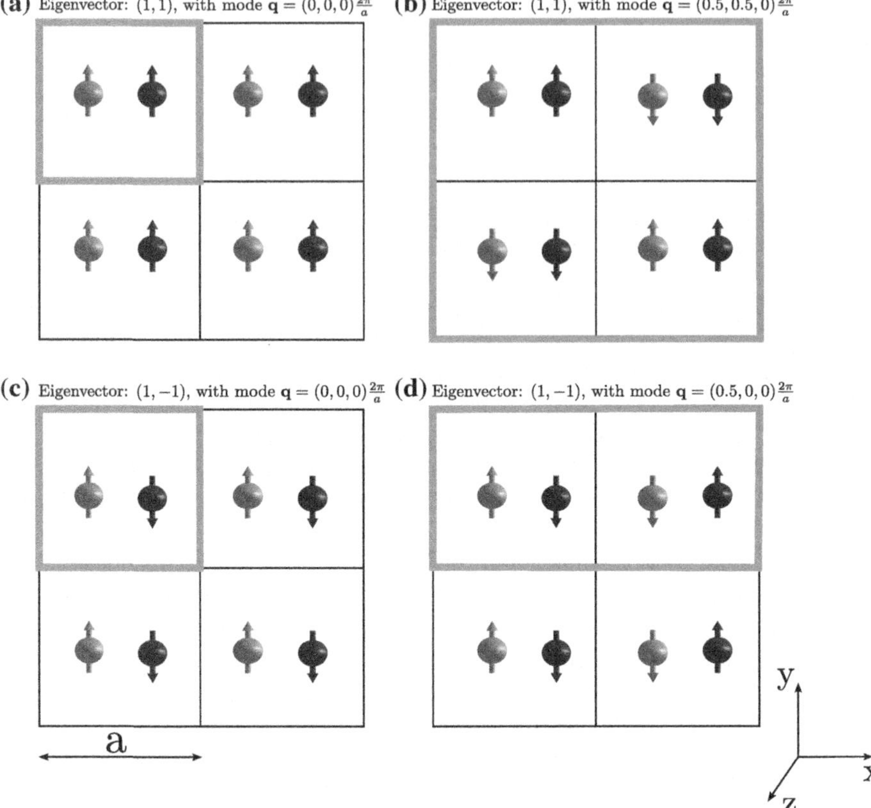

Fig. 3.3 The figure shows four different high temperature magnetic structures of an arbitrary system with two non-equivalent magnetic positions ($N_{\mathrm{at}} = 2$), denoted red and blue (color online), and the corresponding eigenvectors and values of $\mathbf{q}$. The magnetic unit cell is indicated by a thick contour. Note that for the system be ferromagnetic it is necessary that $\mathbf{q} = \mathbf{0}$ and the components of the eigenvector are parallel, as shown in panel (**a**). We also remark that arrows in this figure represent the local order parameters $\{\mathbf{m}_n\}$

function in the paramagnetic limit ($\{\mathbf{m}_n = \mathbf{0}\}$) to the lowest order

$$S_{ij}^{(2)} = -\frac{1}{\pi}\mathrm{Im} \int_{-\infty}^{E_{\mathrm{F}}} dE f(E - \nu) \int d\hat{e}_i \frac{3\hat{e}_i}{4\pi} \frac{\partial}{\partial \mathbf{m}_j} \log \det \left(\underline{D}_i^0 + \delta\underline{D}_i(\hat{e}_i)\right), \quad (3.81)$$

where

$$\underline{D}_i^0 = \begin{pmatrix} \underline{D}_{i+}^0 & \underline{0} \\ \underline{0} & \underline{D}_{i-}^0 \end{pmatrix} = \begin{pmatrix} \underline{1} + (\underline{t}_{i+}^{-1} - \underline{t}_{c\text{-}0,i}^{-1})\underline{\tau}_{c\text{-}0,ii} & \underline{0} \\ \underline{0} & \underline{1} + (\underline{t}_{i-}^{-1} - \underline{t}_{c\text{-}0,i}^{-1})\underline{\tau}_{c\text{-}0,ii} \end{pmatrix}^{-1} \quad (3.82)$$

is the impurity matrix in the paramagnetic state and $\delta \underline{D}_i(\hat{e}_i)$ is its first order correction. Note that we have expressed the matrices in terms of their components in the spin frame of reference in which they are diagonal, i.e. in which the $\hat{z}$-axis is parallel to $\hat{e}_i$ (see Eq. (3.48)). For example,

$$\underline{t}_i(\hat{e}_i) = \frac{1}{2}\left(\underline{t}_{i+}^{-1} + \underline{t}_{i-}^{-1}\right)\underline{I}_2 + \frac{1}{2}\left(\underline{t}_{i+}^{-1} - \underline{t}_{i-}^{-1}\right)\boldsymbol{\sigma}\cdot\hat{e}_i, \tag{3.83}$$

and

$$\underline{D}_i^0(\hat{e}_i) = \frac{1}{2}\left(\underline{D}_{i+}^0 + \underline{D}_{i-}^0\right)\underline{I}_2 + \frac{1}{2}\left(\underline{D}_{i+}^0 - \underline{D}_{i-}^0\right)\boldsymbol{\sigma}\cdot\hat{e}_i, \tag{3.84}$$

while for the effective medium we write

$$\begin{aligned}
\left(\underline{t}_{c,i}^0\right)^{-1} &= \frac{1}{2}\left(\left(\underline{t}_{c,i+}^0\right)^{-1} + \left(\underline{t}_{c,i-}^0\right)^{-1}\right)\underline{I}_2 \\
&+ \frac{1}{2}\left(\left(\underline{t}_{c,i+}^0\right)^{-1} - \left(\underline{t}_{c,i-}^0\right)^{-1}\right)\boldsymbol{\sigma}\cdot\hat{m}_i = \underline{t}_{c\text{-}0,i}^{-1}\underline{I}_2,
\end{aligned} \tag{3.85}$$

and

$$\underline{\tau}_{c,ii}^0 = \frac{1}{2}\left(\underline{\tau}_{c,ii+}^0 + \underline{\tau}_{c,ii-}^0\right)\underline{I}_2 + \frac{1}{2}\left(\underline{\tau}_{c,ii+}^0 - \underline{\tau}_{c,ii-}^0\right)\boldsymbol{\sigma}\cdot\hat{m}_i = \underline{\tau}_{c\text{-}0,ii}\underline{I}_2, \tag{3.86}$$

where we have used the definitions

$$\underline{t}_{c\text{-}0,i}^{-1} \equiv \left(\underline{t}_{c,i+}^0\right)^{-1} = \left(\underline{t}_{c,i-}^0\right)^{-1}, \tag{3.87}$$

$$\underline{\tau}_{c\text{-}0,ii} \equiv \underline{\tau}_{c,ii+}^0 = \underline{\tau}_{c,ii-}^0, \tag{3.88}$$

which exploit the spherical symmetry of $P_n(\hat{e}_n) \approx 1/4\pi$. We explicitly drop the dependence of $\underline{t}_{i\pm}^{-1}$ on $\{\mathbf{m}_n\}$ as we assume the presence of good local moments. Carrying out very simple algebra we can express Eq. (3.81) as[3]

$$\begin{aligned}
S_{ij}^{(2)} &= -\frac{1}{\pi}\text{Im}\int_{-\infty}^{E_{\mathrm{F}}} dE f(E-\nu)\int d\hat{e}_i \frac{3\hat{e}_i}{4\pi}\frac{\partial}{\partial\mathbf{m}_j}\log\left[1 + \text{Tr}\left(\left(\underline{D}_i^0\right)^{-1}\delta\underline{D}_i(\hat{e}_i)\right)\right] \\
&\approx -\frac{1}{\pi}\text{Im}\int_{-\infty}^{E_{\mathrm{F}}} dE f(E-\nu)\int d\hat{e}_i \frac{3\hat{e}_i}{4\pi}\frac{\partial}{\partial\mathbf{m}_j}\text{Tr}\left(\left(\underline{D}_i^0\right)^{-1}\delta\underline{D}_i(\hat{e}_i)\right),
\end{aligned} \tag{3.89}$$

The next step follows by noting that we can write $(\underline{t}_{c,i\pm}^0)^{-1} \approx \underline{t}_{c\text{-}0,i}^{-1} \pm \delta\underline{t}_{c,i}^{-1}$ and $\underline{\tau}_{c,ii\pm}^0 \approx \underline{\tau}_{c\text{-}0,ii}^0 \pm \delta\tau_{0,ii}$, which gives

$$\underline{D}_i \approx \underline{D}_i^0 + \underline{D}_i^0\left[\left(\Delta_i^0 + \delta\Delta_i^0\right)\left(\underline{\tau}_{c\text{-}0,ii} + \delta\underline{\tau}_{c,ii}\boldsymbol{\sigma}\cdot\hat{m}_i\right) - \Delta_i^0\underline{\tau}_{c,ii}^0\right]\underline{D}_i^0, \tag{3.90}$$

---

[3]In this step we have used the equality $\det(A+B) = \det A + \det B + \det B\,\text{Tr}(AB^{-1})$ and neglected a term that does not contribute to $S_{ij}^{(2)}$.

where

$$\underline{\Delta}_i^0 = \frac{1}{2}\left(\underline{t}_{i+}^{-1} + \underline{t}_{i-}^{-1}\right)\underline{I}_2 + \frac{1}{2}\left(\underline{t}_{i+}^{-1} - \underline{t}_{i-}^{-1}\right)\boldsymbol{\sigma}\cdot\hat{e}_i - \underline{t}_{c\text{-}0,i}^{-1}\underline{I}_2, \tag{3.91}$$

such that

$$\delta\underline{\Delta}_i^0 = -\delta\underline{t}_{c,i}^{-1}\boldsymbol{\sigma}\cdot\hat{m}_i. \tag{3.92}$$

Thus, Eq. (3.89) becomes

$$S_{ij}^{(2)} = -\frac{1}{\pi}\text{Im}\int_{-\infty}^{E_F}dEf(E-\nu)\int d\hat{e}_i\frac{3\hat{e}_i}{4\pi}\frac{\partial}{\partial\mathbf{m}_j}\text{Tr}\left[\underline{D}_i^0\left(\underline{\Delta}_i^0\delta\underline{\tau}_{c,ii}\boldsymbol{\sigma}\cdot\hat{m}_i + \delta\underline{\Delta}_i^0\underline{\tau}_{c,ii}^0\right)\right]. \tag{3.93}$$

Developing now the term to be traced yields

$$\underline{D}_i^0\left(\underline{\Delta}_i^0\delta\underline{\tau}_{c,ii}\boldsymbol{\sigma}\cdot\hat{m}_i + \delta\underline{\Delta}_i^0\underline{\tau}_{c,ii}^0\right) =$$
$$\left[\frac{\underline{t}_{i+}^{-1} - \underline{t}_{i-}^{-1}}{2}\delta\underline{\tau}_{c,ii} + \frac{\underline{D}_{i+}^0 - \underline{D}_{i-}^0}{2}\left(\frac{\underline{t}_{i+}^{-1} + \underline{t}_{i-}^{-1}}{2}\delta\underline{\tau}_{c,ii} - \underline{t}_{c\text{-}0,i}^{-1}\delta\underline{\tau}_{c,ii} - \delta\underline{t}_{c,i}^{-1}\underline{\tau}_{c,ii}^0\right)\right](\hat{m}_i\cdot\hat{e}_i)\underline{I}_2$$
$$+\left[\frac{\underline{t}_{i+}^{-1} + \underline{t}_{i-}^{-1}}{2}\delta\underline{\tau}_{c,ii} - \underline{t}_{c\text{-}0,i}^{-1}\delta\underline{\tau}_{c,ii} - \delta\underline{t}_{c,i}^{-1}\underline{\tau}_{c,ii}^0 + \frac{(\underline{D}_{i+}^0 - \underline{D}_{i-}^0)(\underline{t}_{i+}^{-1} + \underline{t}_{i-}^{-1})}{4}(\hat{m}_i\cdot\hat{e}_i)^2\delta\underline{\tau}_{c,ii}\right]\sigma_z. \tag{3.94}$$

Once the integral over the magnetic orientation is performed only the terms accompanied by $(\hat{m}_i\cdot\hat{e}_i)$ in Eq. (3.94) lead to a non-zero contribution. Introducing this into Eq. (3.93) and tracing over the spin degrees only gives

$$S_{ij}^{(2)} = -\frac{1}{\pi}\text{Im}\int_{-\infty}^{E_F}dEf(E-\nu)\text{Tr}_L\left[-(\underline{D}_{i+}^0 - \underline{D}_{i-}^0)\frac{\partial\underline{t}_{c,i}^{-1}}{\partial\mathbf{m}_j}\underline{\tau}_{c\text{-}0,ii}\right.$$
$$\left.+ \underline{D}_{i+}^0(\underline{t}_{i+}^{-1} - \underline{t}_{c\text{-}0,i}^{-1})\frac{\partial\underline{\tau}_{c,ii}}{\partial\mathbf{m}_j} - \underline{D}_{i-}^0(\underline{t}_{i-}^{-1} - \underline{t}_{c\text{-}0,i}^{-1})\frac{\partial\underline{\tau}_{c,ii}}{\partial\mathbf{m}_j}\right], \tag{3.95}$$

where Tr_L traces over the angular momentum labels. The derivatives of this equation can be conveniently rearranged by invoking Eq. (2.66) to show that[4]

$$\frac{\partial\underline{\tau}_{c,ii}}{\partial\mathbf{m}_j} = -\sum_k\underline{\tau}_{c\text{-}0,ik}\frac{\partial\underline{t}_{c,k}^{-1}}{\partial\mathbf{m}_j}\underline{\tau}_{c\text{-}0,ki}, \tag{3.96}$$

from which we obtain

$$S_{ij}^{(2)} = -\frac{1}{\pi}\text{Im}\int_{-\infty}^{E_F}dEf(E-\nu)\text{Tr}_L\left[(\underline{X}_{i+}^0 - \underline{X}_{i-}^0)\sum_{k\neq i}\underline{\tau}_{c\text{-}0,ik}\frac{\partial\underline{t}_{c,k}^{-1}}{\partial\mathbf{m}_j}\underline{\tau}_{c\text{-}0,ki}\right], \tag{3.97}$$

---

[4]Here we have considered the inverse matrix property $\frac{\partial A^{-1}}{\partial\alpha} = -A^{-1}\frac{\partial A}{\partial\alpha}A^{-1}$ too.

Here we have used that $\underline{X}^0_{i\pm} = \underline{D}^0_{i\pm}(\underline{t}^{-1}_{i\pm} - \underline{t}^{-1}_{c\text{-}0,i})$, as defined in Appendix A. Finally, the lattice Fourier transform, as given in Eq. (3.68), is applied to Eq. (3.97) resulting in the expression

$$\tilde{S}^{(2)}_{ss'}(\mathbf{q}) = -\frac{1}{\pi}\mathrm{Im}\int_{-\infty}^{E_{\mathrm{F}}} dE f(E - \nu)\mathrm{Tr}_L\left[(\underline{X}^0_{s+} - \underline{X}^0_{s-})\right.$$
$$\left. \times \sum_{s''}\frac{1}{V_{\mathrm{BZ}}}\int d\mathbf{k}\underline{\tilde{\tau}}_{c\text{-}0,ss''}(\mathbf{k}+\mathbf{q})\underline{\tilde{\Lambda}}_{s''s'}(\mathbf{q})\underline{\tilde{\tau}}_{c\text{-}0,s''s}(\mathbf{k}) - \underline{\tau}_{c\text{-}0,ss}\underline{\tilde{\Lambda}}_{ss}(\mathbf{q})\underline{\tau}_{c\text{-}0,ss}\right], \tag{3.98}$$

where V_{BZ} is the volume of the Brillouin zone,

$$\underline{\Lambda}_{ij} \equiv \frac{\partial \underline{t}^{-1}_{c,i}}{\partial \mathbf{m}_j}, \tag{3.99}$$

and

$$\underline{\tilde{\Lambda}}_{ss'}(\mathbf{q}) = \frac{1}{2}\left(\underline{X}^0_{s-} - \underline{X}^0_{s+}\right)$$
$$- \underline{X}^0_{s+}\left[\sum_{s''}\frac{1}{V_{\mathrm{BZ}}}\int_{V_{\mathrm{BZ}}}d\mathbf{k}\underline{\tilde{\tau}}_{c\text{-}0,ss''}(\mathbf{k}+\mathbf{q})\underline{\tilde{\Lambda}}_{s''s'}(\mathbf{q})\underline{\tilde{\tau}}_{c\text{-}0,s''s}(\mathbf{k}) - \underline{\tau}_{c\text{-}0,ss}\underline{\tilde{\Lambda}}_{ss}(\mathbf{q})\underline{\tau}_{c\text{-}0,ss}\right]\underline{X}^0_{s-}. \tag{3.100}$$

In order to calculate $\tilde{S}^{(2)}_{ss'}(\mathbf{q})$ it is necessary to solve Eq. (3.100), which is obtained from manipulating the CPA equations expressed to the zeroth and lowest order in $\{\mathbf{m}_n\}$. We give detail of this derivation in Appendix A.

References

1. Klenin MA, Hertz JA (1976) Magnetic fluctuations in singlet-ground-state systems. Phys Rev B 14:3024–3035
2. Korenman V, Murray JL, Prange RE (1977) Local-band theory of itinerant ferromagnetism. I. Fermi-liquid theory. Phys Rev B 16:4032–4047
3. Hubbard J (1979) The magnetism of iron. Phys Rev B 19:2626–2636
4. Hubbard J (1979) Magnetism of iron. II. Phys Rev B 20:4584–4595
5. Hubbard J (1981) Magnetism of nickel. Phys Rev B 23:5974–5977
6. Gyorffy BL, Pindor AJ, Staunton J, Stocks GM, Winter H (1985) A first-principles theory of ferromagnetic phase transitions in metals. J Phys F: Met Phys 15(6):1337
7. Mermin ND (1965) Thermal properties of the inhomogeneous electron gas. Phys Rev 137:A1441–A1443
8. Tawil RA, Callaway J (1973) Energy bands in ferromagnetic iron. Phys Rev B 7:4242–4252
9. Liechtenstein AI, Katsnelson MI, Antropov VP, Gubanov VA (1987) Local spin density functional approach to the theory of exchange interactions in ferromagnetic metals and alloys. J Magn Magn Mater 67(1):65–74
10. Feynman RP (1955) Slow electrons in a polar crystal. Phys Rev 97:660–665
11. Takahashi M (1981) Generalization of mean-field approximations by the Feynman inequality and application to long-range Ising chain. J Phys Soc Jpn 50(6):1854–1860

12. Staunton JB, Szunyogh L, Buruzs A, Gyorffy BL, Ostanin S, Udvardi L (2006) Temperature dependence of magnetic anisotropy: an ab initio approach. Phys Rev B 74:144411
13. Zabloudil J, Hammerling R, Szunyogh L, Weinberger P (2005) Electron scattering in solid matter. Springer, Berlin

Chapter 4
Minimisation of the Gibbs Free Energy: Magnetic Phase Diagrams and Caloric Effects

Chapters 2 and 3 contain the DFT-DLM theory adopted in this thesis and the entire formalism of how to use KKR-MST in the context of SDFT to describe magnetism at finite temperatures. The central quantities obtained by the theory are the derivatives of $\langle \Omega^{\text{int}} \rangle_0$, namely the internal local fields $\{\mathbf{h}_n^{\text{int}}\}$ and the direct correlation function. Their calculation as a function of the state of magnetic order $\{\mathbf{m}_n\}$ sets the basis of the study of magnetic structures and their stabilisation. This chapter is devoted to show how to obtain and minimise the Gibbs free energy, $\mathcal{G}_1$, from DFT-DLM calculations for given values of the temperature and external fields. Firstly, in Sect. 4.1 we analyse $\langle \Omega^{\text{int}} \rangle_0$ and show the effect of multi-spin, or multi-site, interactions. Their presence have fundamental consequences on the shape of $\mathcal{G}_1$ and on the magnetic phase diagrams, which we show how to construct in Sect. 4.2, and can be at the origin of the existence of tricritical points. The complexity of $\langle \Omega^{\text{int}} \rangle_0$, however, can make the search of the equilibrium state very difficult. Finding a magnetic ordering $\{\mathbf{m}_n\}$ that generates a set of internal fields $\{\mathbf{h}_n^{\text{int}}\}$ that minimise $\mathcal{G}_1$ clearly can be a complicated mathematical problem for non-trivial magnetic structures. Thus, in Sect. 4.2.1 we present a method to find equilibrium states in a self-consistent calculation. The scheme requires of some initial knowledge of the magnetic structures of interest, which is provided by the calculation of the direct correlation function. We illustrate the methodology by showing some case studies in Sects. 4.2.2 and 4.2.3 as well as its application to the well known bcc iron system in Sect. 4.3. Finally, Sect. 4.4 is dedicated to show how the theory can be used to calculate caloric effects, namely adiabatic temperature changes and isothermal entropy changes, at second- and first-order transitions.

4.1 Multi-spin Interactions

As we have explained in Sect. 3.2, $\tilde{\Omega}(\{\hat{e}_n\})$, and therefore $\Omega^{\text{int}}(\{\hat{e}_n\})$, can have a non-trivial dependence on the local moment orientations. The ensemble average $\langle \Omega^{\text{int}} \rangle_0$ can be, consequently, a very complicated function of $\{\mathbf{m}_n\}$ too. We can express $\langle \Omega^{\text{int}} \rangle_0$ in the most general way as a sum of low and high order terms,

© Springer Nature Switzerland AG 2020

E. Mendive Tapia, *Ab initio Theory of Magnetic Ordering*, Springer Theses,
https://doi.org/10.1007/978-3-030-37238-5_4

$$\langle \Omega^{\text{int}} \rangle_0 = \Omega_0 + f^{(2)}\left(\{\mathbf{m}_n\}\right) + f^{(4)}\left(\{\mathbf{m}_n\}\right) + \cdots . \tag{4.1}$$

Here Ω_0 is a constant and $f^{(n)}\left(\{\mathbf{m}_n\}\right)$ is a n-th order function of $\{\mathbf{m}_n\}$. Note that Eq. (4.1) preserves the symmetry $\{\mathbf{m}_n\} \to -\{\mathbf{m}_n\}$ since these quantities change sign under time reversal. If the magnetic state is isotropic then the second and fourth order functions, for example, can be generally expressed as

$$f^{(2)}\left(\{\mathbf{m}_n\}\right) = -\sum_{ij} \mathcal{J}_{ij}\left(\mathbf{m}_i \cdot \mathbf{m}_j\right) \tag{4.2}$$

$$f^{(4)}\left(\{\mathbf{m}_n\}\right) = -\sum_{ij,kl} \mathcal{K}_{ij,kl}\left(\mathbf{m}_i \cdot \mathbf{m}_j\right)\left(\mathbf{m}_k \cdot \mathbf{m}_l\right), \tag{4.3}$$

where $\mathcal{J}_{ij}$ and $\mathcal{K}_{ij,kl}$ are matrices of rank two and four containing the corresponding quadratic and quartic constants of the expansion, respectively. We stress the multi-spin nature of fourth and higher order terms connecting the local order parameters in multiple sites, which are the direct consequence of the complexity of $\Omega^{\text{int}}(\{\hat{e}_n\})$ and its electronic glue origins. To illustrate the effect of their existence, we study the dependence of the internal magnetic fields on $\{\mathbf{m}_n\}$. Using Eq. (3.28) we can express it in terms of the partial derivatives

$$\mathbf{h}_i^{\text{int}} = -\frac{\partial f^{(2)}\left(\{\mathbf{m}_n\}\right)}{\partial \mathbf{m}_i} - \frac{\partial f^{(4)}\left(\{\mathbf{m}_n\}\right)}{\partial \mathbf{m}_i} - \cdots . \tag{4.4}$$

We can now take the limit of approaching the paramagnetic state and write

$$\lim_{\{\mathbf{m}_n \to 0\}} \mathbf{h}_i^{\text{int}} = \sum_j \mathcal{J}_{ij}\mathbf{m}_j . \tag{4.5}$$

Note that from this equation we can use Eq. (3.55) to show that the quadratic constants are nothing else than the direct correlation function in the paramagnetic limit,

$$\lim_{\{\mathbf{m}_n \to 0\}} S_{ij}^{(2)}\left(\{\mathbf{m}_n\}\right) = \mathcal{J}_{ij} . \tag{4.6}$$

Equation (4.5) shows that at temperatures close to the paramagnetic state, when the ordering at every magnetic site is nearly non-existent, the magnetic interactions are well described by pairwise terms. However, as the magnetic ordering increases, by reducing the temperature or applying an external magnetic field for example, the local order parameters increase in size and, consequently, higher order terms become important and must be considered in Eq. (4.1). In other words, the presence of multi-spin (or multi-site) interactions effectively changes the magnetic behaviour as the ordering develops at each site. The meaning of this is that fluctuations of the local moment orientations at different temperatures can induce qualitative changes on the nature of the electronic glue, which spin-polarises and in turn affects the magnetic interactions between the local moments due to its changes. An indication of this has been found by several authors in the past who calculated effective pairwise interactions from the FM state to be different from those in the PM state [1, 2] and some AFM phases [3], as well as significant differences between PM and ferrimagnetic (FIM) states [4].

4.2 Magnetic Phase Diagrams

We construct magnetic phase diagrams by calculating and comparing the Gibbs free energy $\mathcal{G}_1$, introduced in Eq. (3.26), of magnetic structures of interest. The magnetic ordering $\{\mathbf{m}_n\}$ that globally minimises $\mathcal{G}_1$ is considered as the most stable structure at every point in the diagram, defined at different values of the temperature and other parameters, such as the strength of an external magnetic field for example. We point out, however, that our DFT-DLM machinery is not prepared to directly calculate $\langle \Omega^{\text{int}} \rangle_0$, which is necessary to obtain $\mathcal{G}_1$, but its first and second derivatives, $\{\mathbf{h}_n^{\text{int}}\}$ and the direct correlation function. We find the dependence of $\langle \Omega^{\text{int}} \rangle_0$ on the magnetic ordering by thoroughly sampling $\{\mathbf{h}_n^{\text{int}}\}$ against the extensive space $\{\mathbf{m}_n\}$ around the magnetic structures of interest. The data obtained then can be used to evaluate Eq. (4.4) and, hence, extract the pairwise and higher order constants $\mathcal{J}_{ij}$, $\mathcal{K}_{ij,kl}$, etc that accurately describes $\{\mathbf{h}_n^{\text{int}}\}$ numerically. Once these constants are obtained, $\langle \Omega^{\text{int}} \rangle_0$, and consequently $\mathcal{G}_1$, are accessible and can be computed from Eqs. (3.26) and (4.1), respectively. This is the procedure to calculate $\mathcal{G}_1$ at different temperatures and values of an external magnetic field. In addition, by repeating the examination of $\{\mathbf{h}_n^{\text{int}}\}$ at different lattice spacings and interatomic distances the dependence of the pairwise and multi-site interactions on the lattice structure can be obtained too.

Since the shape and minima location of $\mathcal{G}_1$ can be monitored by this methodology, the theory is suitable to characterise the order of phase transitions as well as to predict tricritical points. As shown by Bean and Rodbell in their study of the ferromagnetic-paramagnetic phase transition in MnAs [5], first-order transitions are originated by the presence of an overall negative contribution from the fourth order in the order parameter to the free energy. It is instructive to expand the single-site entropy defined in Eq. (3.24) in terms of its natural dependence on $\lambda_n = \beta h_n$,

$$S_n = k_{\text{B}} \left(\log 4\pi - \frac{1}{6}(\beta h_n)^2 + \frac{1}{60}(\beta h_n)^4 - \frac{1}{567}(\beta h_n)^6 + \cdots \right). \tag{4.7}$$

It is a matter of also expanding the local order parameters in Eq. (3.18) and rearranging terms to find from the previous equation that

$$S_n = k_{\text{B}} \left(\log 4\pi - \frac{3}{2}m_n^2 - \frac{9}{20}m_n^4 - \frac{99}{350}m_n^6 - \cdots \right). \tag{4.8}$$

In the Bean and Rodbell model [5] the first-order character of the transition is driven by a negative quartic term generated by the presence of a magnetovolume coupling.

The theory presented here, however, is also able to quantify the four-site constants $\mathcal{K}_{ij.kl}$ and identify if their presence can be at the origin of the order character in specific phase transitions under study. We emphasise that transitions between ordered-to-ordered magnetic phases can be predicted too if the magnetic dependence in Eq. (4.1) generates them.

To exemplify the free energy minimisation, in Sects. 4.2.2 and 4.2.3 we show two hypothetical but illustrative cases and study the effect of the pairwise and four-site interactions. Before proceeding with them a method to find Gibbs free energy minima self-consistently is firstly introduced in Sect. 4.2.1.

4.2.1 The Equilibrium State: A Self-consistent Calculation

The first step is to identify the most stable magnetic structures at high temperatures by exploring the instabilities of the paramagnetic state, following the calculation described in Sect. 3.4. Once we know the magnetic arrangements of interest we construct the appropriate magnetic unit cell to describe them. Note that a magnetic ordering is defined by the local order parameters $\{\mathbf{m}_n\}$ assigned to each magnetic site, which in turn are prescribed by $\{\boldsymbol{\lambda}_n = \beta \mathbf{h}_n\}$. For example, in a ferromagnetic phase the local moments at every magnetic site fluctuate such that all associated local order parameters have identical sizes and point along the same direction. Thus, the magnetic unit cell in this case contains only one non-equivalent magnetic position and one order parameter $\mathbf{m}_{\mathrm{FM}}(\boldsymbol{\lambda}_{\mathrm{FM}})$ is enough to describe the ferromagnetic ordering. Moreover, one can consider a magnetic system that orders antiferromagnetically via the anti-parallel spin-polarisation of two sub-lattices. Then two order parameters $\{\mathbf{m}_{\mathrm{up}}(\boldsymbol{\lambda}_{\mathrm{up}}), \mathbf{m}_{\mathrm{down}}(\boldsymbol{\lambda}_{\mathrm{down}})\}$ pointing in opposite directions are in principle necessary. We remark that different types of atoms and/or non-equivalent lattice positions naturally develop different magnetic ordering, i.e. they have different values of the local order parameters. Supposing that N is the number of non-equivalent magnetic positions inside the magnetic unit cell, we have that the magnetic ordering is unequivocally described by $\{\mathbf{m}_1(\boldsymbol{\lambda}_1), \mathbf{m}_2(\boldsymbol{\lambda}_2), \ldots, \mathbf{m}_N(\boldsymbol{\lambda}_N)\}$.

Our disordered local moment theory is designed to calculate the emerging internal fields when the electronic structure is constrained to a specific magnetic arrangement described by the single-site probabilities $\{P_n(\hat{e}_n)\}$, which are prescribed by the quantities $\{\boldsymbol{\lambda}_n = \beta \mathbf{h}_n\}$ or equivalently $\{\mathbf{m}_n\}$. The equilibrium state of the N local order parameters is obtained at temperature T and external magnetic field $\mathbf{H}$ when Eq. (3.29) is satisfied at every site, i.e. $\{\mathbf{h}_n = \mathbf{h}_n^{\mathrm{int}} + \mu_n \mathbf{H}\}$. This condition can be written as

$$k_{\mathrm{B}} T = \frac{\mathbf{h}_1^{\mathrm{int}} + \mu_1 \mathbf{H}}{\boldsymbol{\lambda}_1} = \frac{\mathbf{h}_2^{\mathrm{int}} + \mu_2 \mathbf{H}}{\boldsymbol{\lambda}_2} = \cdots = \frac{\mathbf{h}_N^{\mathrm{int}} + \mu_N \mathbf{H}}{\boldsymbol{\lambda}_N}. \tag{4.9}$$

Finding the magnetic configuration $\{\boldsymbol{\lambda}_1, \ldots, \boldsymbol{\lambda}_N\}$ that generates the appropriate fields $\{\mathbf{h}_1^{\mathrm{int}}, \ldots, \mathbf{h}_N^{\mathrm{int}}\}$ satisfying Eq. (4.9) is something that might not be found directly except in trivial cases such as in the ferromagnetic phase. For given values of the temperature T and external magnetic field $\mathbf{H}$, we solve self-consistently this set of equations by the following procedure

(i) We make an initial guess $\{\mathbf{m}_1^0(\boldsymbol{\lambda}_1^0), \ldots, \mathbf{m}_N^0(\boldsymbol{\lambda}_N^0)\}$ that approximates closely to the magnetic configuration of interest and find the corresponding quantities $\{\boldsymbol{\lambda}_1^0, \ldots, \boldsymbol{\lambda}_N^0\}$ by solving Eq. (3.18).

(ii) We proceed to perform the DFT-based DLM calculation of the internal magnetic fields to obtain $\{\mathbf{h}_1^{\text{int0}} + \mu_1\mathbf{H}, \ldots, \mathbf{h}_N^{\text{int0}} + \mu_N\mathbf{H}\}$ by constraining the underlying electronic structure to the probability distribution $P(\{\hat{e}_n\})$ prescribed by $\{\boldsymbol{\lambda}_1^0, \ldots, \boldsymbol{\lambda}_N^0\}$ and using Eq. (3.53). Assuming the presence of good local moments rules out the necessity to recalculate self-consistently the potentials for different magnetic orderings.

(iii) We check if the self-consistent condition in Eq. (4.9) is satisfied. If not, we make a new guess composed by the initial one, $\{\boldsymbol{\lambda}_1^0, \ldots, \boldsymbol{\lambda}_N^0\}$, and the output from step (ii) as

$$\{\boldsymbol{\lambda}_1^0, \ldots, \boldsymbol{\lambda}_N^0\}_{\text{new}} = c_1\{\boldsymbol{\lambda}_1^0, \ldots, \boldsymbol{\lambda}_N^0\} + c_2\frac{1}{k_{\text{B}}T}\{\mathbf{h}_1^{\text{int0}} + \mu_1\mathbf{H}, \ldots, \mathbf{h}_N^{\text{int0}} + \mu_N\mathbf{H}\}$$

$$(4.10)$$

where c_1 and c_2 are some mixing constants.

(iv) Steps (ii) and (iii) are repeated iteratively until the condition in Eq. (4.9) is satisfied.

4.2.2 Case Study: One Type of Magnetic Ordering

Here we consider the simple situation in which there is interest in one type of magnetic ordering only under the presence of no external magnetic field. We also assume that even if the local order parameters $\{\mathbf{m}_n\}$ defining the magnetic ordering form a complicated magnetic structure, their sizes are equal ($m_n = m$ for all n) and their directions do not change with temperature. For example, the simplest system matching these conditions would be a ferromagnet with only one non-equivalent magnetic position, but other more complicated non-collinear structures can be described by a single local order parameter too. Under these prescriptions, the quadratic and quartic constants can be compactly accounted with

$$a = N \sum_{ij} \mathcal{J}_{ij} \cos\left(\theta_{ij}\right) \tag{4.11}$$

$$b = N \sum_{ij,kl} \mathcal{K}_{ij,kl} \cos\left(\theta_{ij}\right) \cos\left(\theta_{kl}\right), \tag{4.12}$$

where θ_{ij} is the angle between $\mathbf{m}_i$ and $\mathbf{m}_j$ and N is the number of magnetic sites inside the magnetic unit cell. Recalling Eqs. (3.26), (3.25), and (4.1), the free energy per magnetic atom is then given by

$$\mathcal{G}_{1,n} = -am^2 - bm^4 - T S_n(\beta h_n) \tag{4.13}$$

and the internal field as

$$\mathbf{h}_n^{\mathrm{int}} = (2am + 4bm^3)\hat{m}_n. \tag{4.14}$$

Evidently, the equilibrium condition is simply $\{\mathbf{h}_n^{\mathrm{int}} = \mathbf{h}_n\}$. If the transition is continuous the temperature at which it occurs is readily obtained by $\nabla_{m_n}\mathcal{G}_{1,n} = 0$ in the limit of $\{\mathbf{m}_n\} \to \mathbf{0}$. Using the entropy expansion given in Eq. (4.8) this results in

$$T_{tr(2nd)} = \frac{2a}{3k_{\mathrm{B}}}. \tag{4.15}$$

Note that this is Eq. (3.76) expressed in terms of pairwise constants, compactly contained inside a. The critical value of b from which the transition becomes first-order is found by requiring that the fourth order is zero at $T = T_{tr(2nd)}$, which gives

$$b = \frac{3a}{10}. \tag{4.16}$$

This means that for $b < 3a/10$ the phase transition is continuous and for $b > 3a/10$ it is discontinuous. In order to calculate equilibrium states it is a matter of standard numerical minimisation of $\mathcal{G}_{1,n}$ with respect to m_n at different temperatures and model parameters a and b. In Fig. 4.1 we summarise the results. We suggest the reader to pay special attention to panels (c)–(e), where the behaviour of the free energy against the local order parameter is shown, illustrating the characterisation of the order of the transition. A first-order transition is described by a sudden change of the lowest minimum location, occurring at $T = T_{Tr}$ in panel (e). The critical point separating first- and second-order transitions is indicated in panel (a).

4.2.3   Case Study: Ferromagnetism Versus Antiferromagnetism

In this case we study a magnetic unit cell with two identical magnetic atoms of the same type and at equivalent magnetic positions. The aim is to show the stabilisation competition between the ferromagnetic phase, in which the local order parameters associated with the two sites are equal and parallel, $\mathbf{m}_1 = \mathbf{m}_2$, and the antiferromagnetic phase, in which they are fully anti-parallel, $\mathbf{m}_1 = -\mathbf{m}_2$. To study this we consider that the free energy per magnetic unit cell is well described by

$$\begin{aligned}
\mathcal{G}_1 = &- a(m_1^2 + m_2^2) - b(m_1^4 + m_2^4) - \alpha\mathbf{m}_1 \cdot \mathbf{m}_2 \\
&+ \frac{1}{2}\gamma(m_1^2 + m_2^2)\mathbf{m}_1 \cdot \mathbf{m}_2 - \gamma_2 m_1^2 m_2^2 - T(S_1(\beta h_2) + S_2(\beta h_2)).
\end{aligned} \tag{4.17}$$

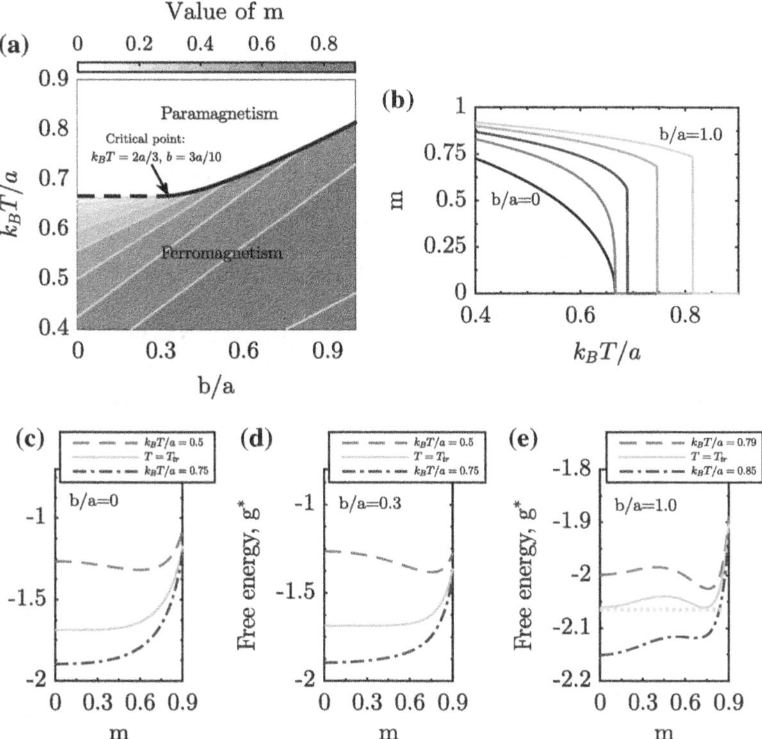

Fig. 4.1 a The temperature—b/a phase diagram. Dashed and continuous black lines indicate second- and first-order transitions, respectively. **b** m against temperature for pertinent values of the model parameters above and below the critical point. **c, d, e** The free energy below, at, and above the transition temperature for relevant model parameters, showing the difference between second- and first- order transitions

Similarly to the first case study, due to the symmetry of the problem the pair- and four- site interactions can be compactly comprised in a small set of constants, $\{a, \alpha\}$ and $\{b, \gamma, \gamma_2\}$, respectively. Taking the derivatives with respect to $\mathbf{m}_1$ and $\mathbf{m}_2$ yields internal magnetic fields

$$\mathbf{h}_1^{\text{int}} = 2a\mathbf{m}_1 + 4bm_1^2\mathbf{m}_1 + \alpha\mathbf{m}_2 - \frac{1}{2}\gamma\left[2(\mathbf{m}_1 \cdot \mathbf{m}_2)\mathbf{m}_1 + (m_1^2 + m_2^2)\mathbf{m}_2\right], \quad (4.18)$$

$$\mathbf{h}_2^{\text{int}} = 2a\mathbf{m}_2 + 4bm_2^2\mathbf{m}_2 + \alpha\mathbf{m}_1 - \frac{1}{2}\gamma\left[2(\mathbf{m}_1 \cdot \mathbf{m}_2)\mathbf{m}_2 + (m_1^2 + m_2^2)\mathbf{m}_1\right]. \quad (4.19)$$

Here we merely study the effect of the constants $\{a, b, \alpha, \gamma, \gamma_2\}$, but the calculation of the internal fields must be performed from the DFT-DLM machinery to obtain ab initio $\mathcal{J}_{ij}$ and $\mathcal{J}_{ij,kl}$. Again, second-order transition temperatures to both ferromagnetic and antiferromagnetic states can be obtained from the expansion given in Eq. (4.8) and taking the paramagnetic limit of $\nabla_{\mathbf{m}_{1(2)}}\mathcal{G}_1 = \mathbf{0}$, which gives

$$\mathbf{m}_1 = \mathbf{m}_2 \Rightarrow T_{tr(2nd,F)} = \frac{2a + \alpha}{3k_{\mathrm{B}}}, \tag{4.20}$$

$$\mathbf{m}_1 = -\mathbf{m}_2 \Rightarrow T_{tr(2nd,AF)} = \frac{2a - \alpha}{3k_{\mathrm{B}}}. \tag{4.21}$$

Thus, for positive values of α the paramagnetic state becomes unstable to the formation of ferromagnetic ordering. Moreover, by inspecting Eq. (4.17) it is evident that positive values of γ favour antiferromagnetism. To reveal this effect we set $b = 0$, $\gamma_2 = 0$ and $\alpha/a = +0.2$, and minimise $\mathcal{G}_1$ at different temperatures and values of γ. As shown in Fig. 4.2a, the phase diagram presents two tricritical points. (A) corresponds to the critical regime in which the system at $T = 0$ is energetically equal for both magnetic phases. By defining $g_F^* = \mathcal{G}_1(\mathbf{m}_1 = \mathbf{m}_2)/a$ and $g_{AF}^* = \mathcal{G}_1(\mathbf{m}_1 = -\mathbf{m}_2)/a$, it follows that $g_F^* = g_{AF}^*$ at the critical point (A), which is therefore defined as $\gamma = \alpha$. On the other hand, for high enough values of γ the fourth order coupling is able to stabilise antiferromagnetism above $T_{tr(2nd,F)}$ and a tricritical point (B) exists in consequence.

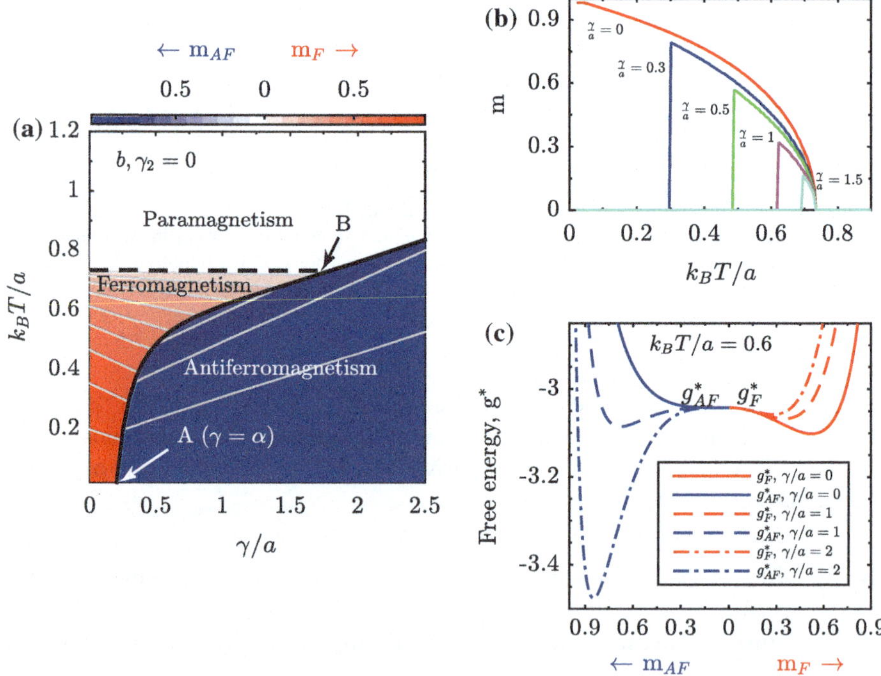

Fig. 4.2 a The temperature—γ/a phase diagram with the colour scheme defined as $\mathbf{m}_F = \mathbf{m}_1 = \mathbf{m}_2$ in red when $g_F^* < g_{AF}^*$ and $\mathbf{m}_{AF} = \mathbf{m}_1 = -\mathbf{m}_2$ in blue when $g_F^* > g_{AF}^*$. Dashed and continuous black lines indicate second- and first-order transitions, respectively. **b** The total order parameter normalised inside the unit cell, $\mathbf{m} = (\mathbf{m}_1 + \mathbf{m}_2)/2$, for increasing values of γ/a. **c** The free energies g_F^* (red and right side) and g_{AF}^* (blue and left side) for pertinent values of γ/a illustrating the first-order behaviour

4.3   Application to bcc Iron

In this section we perform DFT-DLM calculations to study the well known bcc iron system. This compound has a lattice parameter of 2.86 Åand presents a continuous ferromagnetic-paramagnetic phase transition at $T^{\text{exp}} = 1044$ K. Its magnetism, therefore, can be described by one local order parameter $\mathbf{m}_{Fe}$ associated with the Fe atoms in a bcc magnetic unit cell. Consequently, $N = 1$ and the lattice Fourier transform of the direct correlation function in the limit of the paramagnetic state contains only one component which is equal to its single eigenfunction, $\tilde{S}_{s=1s'=1}(\mathbf{q}) = \tilde{u}_{\text{P}}(\mathbf{q})$.

Firstly, we obtained the fully disordered paramagnetic potentials by using our DFT-DLM machinery in a scalar-relativistic calculation of bcc iron with a lattice parameter of $a = 2.79$ Å. Then these potentials were used to calculate $\tilde{u}_{\text{P}}(\mathbf{q})$, which we show in Fig. 4.3. As expected, this function peaks at $\mathbf{q}_{\text{P}} = \mathbf{0}$ favouring the stabilisation of the ferromagnetic state at high temperatures. Note that the calculation does not show any trait of a stabilisation of another more complicated magnetic structure. Since we are interested in describing the ferromagnetic phase only, following our methodology in Sect. 4.2 we write the internal energy $\langle \Omega^{\text{int}} \rangle_0$ as an expansion in terms of $\mathbf{m}_{Fe}$

$$\langle \Omega^{\text{int}} \rangle_0 = -am_{Fe}^2 - bm_{Fe}^4 - cm_{Fe}^6 - \cdots . \tag{4.22}$$

We then proceed to evaluate the emerging field $\mathbf{h}_{Fe}^{\text{int}}$ against different values of $\mathbf{m}_{Fe}$, i.e., we repeat the calculation for various probability distributions prescribed by $\mathbf{m}_{Fe}$ using the fully disordered paramagnetic potentials. We show the results in Fig. 4.4a. As it can be appreciated from the figure, we found that the dependence is nearly linear such that two parameters in the expansion were sufficient to fit our data, $a = 172$ meV and $b = -2$ meV. The fitting curves are shown by continuous lines. The fact that the quartic coefficient is negligible shows that the feedback from the underlying electronic structure when the ferromagnetic ordering increases does not effectively influence the magnetic interactions, yielding a simple second-order behaviour of

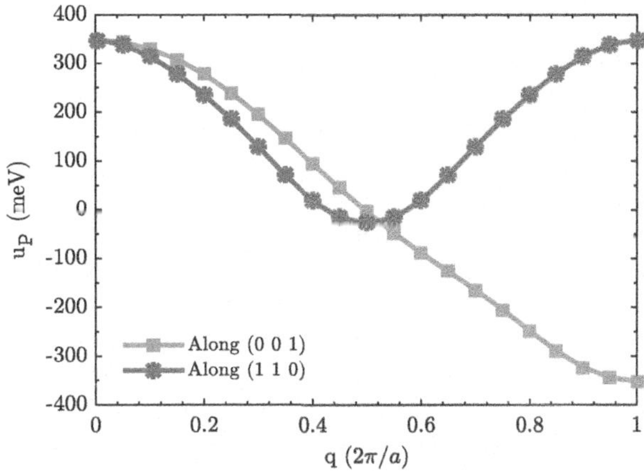

Fig. 4.3 The lattice Fourier transform of the direct correlation function, $\tilde{S}_{s=1s'=1}(\mathbf{q}) = \tilde{u}_{\text{P}}(\mathbf{q})$, against the wave vector $\mathbf{q}$ at the paramagnetic state. The figure shows two characteristic directions of $\mathbf{q}$ along (0 0 1) and (1 1 0)

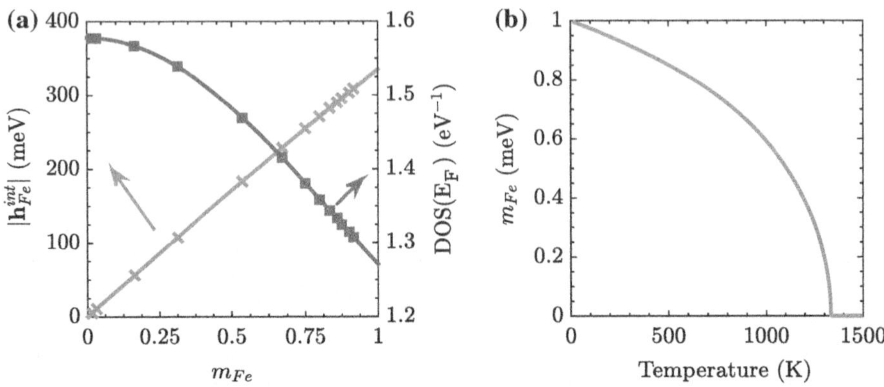

Fig. 4.4 **a** The field $\mathbf{h}^{int}_{Fe}$ (blue squares) and the density of states at the Fermi energy (red crosses) against the ferromagnetic order parameter $\mathbf{m}_{Fe}$. Straight lines are the fitting functions of these quantities. **b** The temperature dependence of the local order parameter

the magnetisation, as shown in Fig. 4.4b. The transition temperature obtained from our calculation is T_c=1330 K, slightly above the experimentally observed, which is expected from a mean-field theory. We point out that the fact that all local order parameters at every magnetic site are the same makes the equilibrium condition expressed in Eq. (4.9) to be trivially satisfied for each pair of values of $\mathbf{h}^{int}_{Fe}$ and $\mathbf{m}_{Fe}$. The corresponding temperature at each point is $T = \frac{\mathbf{h}^{int}_{Fe}}{k_B \lambda_{Fe}}$, where $\lambda_{Fe}(\mathbf{m}_{Fe})$ is calculated from Eq. (3.18). For completeness, we show in panel (a) of the same figure the density of states at the Fermi level as a function of $\mathbf{m}_{Fe}$. It presents a parabolic dependence which shows that the magnetic ordering has an influence on the electronic structure.

4.4 Caloric Effects

Caloric responsive materials are systems that show a substantial change of their thermodynamic state when an external field is applied and/or removed on them [6–8]. Depending on what external stimuli triggers the change the effect is called magnetocaloric (MCE) [9], barocaloric (BCE) [10], electrocaloric (ECE) [11, 12], elastocaloric (eCE) [13] and toroidocaloric (TCE) [14–17] effect, for an external influence being a magnetic field, a hydrostatic pressure, an electric field, an uniaxial/biaxial stress and a toroidic field, respectively. Note that the BCE and eCE effects are particular cases of responses driven by mechanical stresses. For this reason they are usually referred to as mechanocaloric effects in general. In 1997 Pecharsky and Gschneidner published what today is considered the work representing a breakthrough on the research of magnetic refrigeration [9]. They showed that materials exhibiting giant caloric responses can also be found at room temperature and used as refrigerants. Magnetic refrigeration based on the exploitation of one or multiple caloric effects has become a widely investigated technology and promises to be a environmentally friendly and more efficient alternative to gas-compression based devices. The interest is to induce large thermodynamical changes as a response to small or moderate external influences, which are most likely to occur in the vicinity of phase transitions, especially if they are of first-order. The magnetic refrigeration community, however, has little guidance from theoretical research and its scientific advances are often heuristic in nature. The disordered local moment theory presented here is designed to naturally evaluate from first-principles entropy changes at high temperatures by evaluating the thermodynamic change of the magnetic ordered state and the consequent effect on the underlying electronic properties. In addition, the capability of the approach to characterise the order of the transitions as well as to predict and locate tricritical points make the theory suitable to identify the best cooling cycles from the constructed phase diagrams.

In general, a caloric effect is quantified by the isothermal entropy change, ΔS_{iso}, and the adiabatic temperature change, ΔT_{ad}, induced in the thermodynamic conjugate of the external field applied and/or removed. Using fundamental thermodynamics and the appropriate Maxwell relations, the entropy change of the system when a given external field (Y) is modified isothermally is given by

$$\Delta S_{iso}(T, 0 \rightarrow Y) = S(Y, T) - S(0, T) = \int_0^Y \left(\frac{\partial S}{\partial Y}\right)_T dY = \int_0^Y \left(\frac{\partial X}{\partial T}\right)_Y dY,$$
(4.23)

where X is the thermodynamical conjugate of Y. On the other hand, the adiabatic temperature change is expressed as

$$\Delta T_{ad}(T, 0 \rightarrow Y) = -\int_0^Y \frac{T}{C}\left(\frac{\partial S}{\partial Y}\right)_T dY = -\int_0^Y \frac{T}{C}\left(\frac{\partial X}{\partial T}\right)_Y dY, \qquad (4.24)$$

C being the heat capacity. Instead of performing these integrals our strategy consists in exploiting the fact that our theory can directly provide the total entropy as a function of the state of magnetic order and, therefore, against temperature, magnetic field and lattice spacing. It naturally predicts the entropy changes due to the orientational disorder of the local moments, ΔS_{mag}, from Eqs. (3.23) and (3.24). Moreover, we can also estimate the entropy change from thermal alterations of the density of states by using the Sommerfeld expansion leading to [18]

$$S_{elec} = \frac{\pi^2}{3}k_B^2 T n(E_F), \qquad (4.25)$$

where $n(E_F)$ is the electronic density of states at the Fermi energy. We name this contribution electronic entropy and due to the nature of the time scale separation of the disordered local moment approach it is formally captured within the internal energy

$$\langle \Omega^{int} \rangle_0 = \bar{E} - T S_{elec}, \qquad (4.26)$$

where $\bar{E}$ is the DFT-based energy averaged over local moment configurations [19, 20]. Note that owing that $n(E_F)$, and consequently S_{elec}, depend on the state of magnetic order, Eq. (4.26) can be used to partly elucidate the source of entropy change at magnetic phase transitions by calculating the entropy at different magnetic states. Moreover, following the Born-Oppenheimer approximation it would be reasonable to consider the lattice vibrations to fluctuate in the slowest time scale $\tau_{\text{vib}} \gg \tau_{\text{mag}} \gg \tau_{\text{elec}}$. Under these circumstances it should be possible to use the same ideas developed for the magnetic fluctuations and expand the theory to incorporate the effect of the vibrational fluctuations. Such a theory should be able to address the entire magneto-phonon coupling at finite temperatures, providing that the time-scale assumptions are correct. At the current state of the theory the atomic positions are not allowed to move and such an effect is not taken into account. We follow an alternative route and incorporate the effect of the lattice vibrations via the implementation of a simple Debye model defining the corresponding entropy as [21]

$$S_{vib} = k_{\text{B}}\left[-3\ln\left(1 - e^{-\frac{\theta_{\text{D}}}{T}}\right) + 12\left(\frac{T}{\theta_{\text{D}}}\right)^3 \int_0^{\frac{\theta_{\text{D}}}{T}} \frac{x^3}{e^x - 1}dx\right], \qquad (4.27)$$

where θ_{D} is the Debye temperature, which can be obtained from experiment or other sources. The presence of this term aims only to purely act as a thermal bath or reservoir to exchange entropy with the electronic and magnetic degrees of freedom. Its inclusion is necessary to avoid non-physical results when calculating ΔT_{ad} [22]. Hence, in our approach the total entropy is directly calculated as

$$S_{tot} = S_{\text{mag}} + S_{elec} + S_{vib}. \qquad (4.28)$$

This is the central equation in our method for the calculation of caloric effects. For example, under the presence of an external magnetic field that varies as $\mathbf{H} = \mathbf{H}_0 \rightarrow \mathbf{H}_1$ we can compute the isothermal entropy change from

$$\Delta S_{iso}(T, \mathbf{H}_0 \rightarrow \mathbf{H}_1) = S_{tot}(\mathbf{H}_1, T) - S_{tot}(\mathbf{H}_0, T). \qquad (4.29)$$

Similarly, we obtain the adiabatic temperature change by solving the equation

$$S_{tot}(T, \mathbf{H}_0) = S_{tot}(T + \Delta T_{ad}, \mathbf{H}_1). \qquad (4.30)$$

In addition, mechanocaloric effects can be estimated by the calculation of thermal responses caused by the change of lattice spacings, that is

$$\Delta S_{iso}(T, \varepsilon_0 \rightarrow \varepsilon_1) = S_{tot}(\varepsilon_1, T) - S_{tot}(\varepsilon_0, T), \qquad (4.31)$$

$$S_{tot}(T, \varepsilon_0) = S_{tot}(T + \Delta T_{ad}, \varepsilon_1), \qquad (4.32)$$

where ε_0 and ε_1 represent the lattice parameters and/or atomic positions before and after the stress application, respectively, either biaxial, uniaxial, or isotropic.

Evidently, from these equations the theory is also ready to calculate the field-tune and multicaloric effects involving the magneto- and mechano- caloric responses. In the first situation, a secondary field is kept constant during the variation of the primary field. While the primary field drives the caloric response, the secondary field plays the role of adjusting the best operative conditions. Moreover, the multicaloric effect refers to the variation of the two fields either simultaneously or sequentially. To finalise this section, we point out that both conventional and inverse caloric effects can be modelled by the approach presented. Contrary to conventional caloric responses, the inverse effect is based on cooling by adiabatic magnetisation when an external magnetic field is applied, yielding a positive increment of entropy. This effect is observed in systems showing first-order metamagnetic transitions where the magnetisation shows a sudden raise by increasing the temperature, as it is immediately shown by Eq. (4.23). Some examples in which inverse MCEs effects have been studied are the off-stoichiometry FeRh system [20, 23, 24], the ferrimagnetic-antiferromagnetic transition in doped Mn_2Sb compounds [25, 26], and the non-collinear magnetism in Mn_5Si_3 [27].

References

1. Khmelevskyi S, Khmelevska T, Ruban AV, Mohn P (2007) Magnetic exchange interactions in the paramagnetic state of hcp Gd. J Phys: Condens Matter 19(32):326218
2. Shallcross S, Kissavos AE, Meded V, Ruban AV (2005) An ab initio effective Hamiltonian for magnetism including longitudinal spin fluctuations. Phys Rev B 72:104437
3. Antal A, Lazarovits B, Udvardi L, Szunyogh L, Újfalussy B, Weinberger P (2008) First-principles calculations of spin interactions and the magnetic ground states of Cr trimers on Au (111). Phys Rev B 77:174429
4. Khmelevskyi S, Ruban AV, Mohn P (2016) Magnetic ordering and exchange interactions in structural modifications of Mn_3Ga alloys: interplay of frustration, atomic order, and off-stoichiometry. Phys Rev B 93:184404
5. Bean CP, Rodbell DS (1962) Magnetic disorder as a first-order phase transformation. Phys Rev 126:104–115
6. Tishin AM, Spichkin YI (2003) The magnetocaloric effect and its applications, vol 6
7. Planes A, Mañosa L, Saxena A (2014) Magnetism and structure in functional materials. Springer, Berlin
8. Sandeman KG (2012) Magnetocaloric materials: the search for new systems. Scr Mater 67(6):566–571. Viewpoint set no. 51: magnetic materials for energy
9. Pecharsky VK, Gschneidner KA Jr (1997) Giant magnetocaloric effect in $Gd_5(Si_2Ge_2)$. Phys Rev Lett 78:4494–4497
10. Matsunami D, Fujita A, Takenaka K, Kano M (2015) Giant barocaloric effect enhanced by the frustration of the antiferromagnetic phase in Mn_3GaN. Nat Mater 14:73
11. Neese B, Chu B, Lu S-G, Wang Y, Furman E, Zhang QM (2008) Large electrocaloric effect in ferroelectric polymers near room temperature. Science 321(5890):821–823
12. Moya X, Stern-Taulats E, Crossley S, González-Alonso D, Kar-Narayan S, Planes A, Mañosa L, Mathur ND (2013) Giant electrocaloric strength in single-crystal $BaTiO_3$. Adv Mater 25(9):1360–1365
13. Bonnot E, Romero R, Mañosa L, Vives E, Planes A (2008) Elastocaloric effect associated with the martensitic transition in shape-memory alloys. Phys Rev Lett 100:125901

14. Spaldin NA, Fiebig M, Mostovoy M (2008) The toroidal moment in condensed-matter physics and its relation to the magnetoelectric effect. J Phys: Condens Matter 20(43):434203
15. Castán T, Planes A, Saxena A (2012) Thermodynamics of ferrotoroidic materials: Toroidocaloric effect. Phys Rev B 85:144429
16. Toledano P, Khalyavin DD, Chapon LC (2011) Spontaneous toroidal moment and field-induced magnetotoroidic effects in $Ba_2CoGe_2O_7$. Phys Rev B 84:094421
17. Baum M, Schmalzl K, Steffens P, Hiess A, Regnault LP, Meven M, Becker P, Bohatý L, Braden M (2013) Controlling toroidal moments by crossed electric and magnetic fields. Phys Rev B 88:024414
18. Ashcroft NW, Mermin DN (1976) Solid state physics. Saunders College Publishing, Philadelphia
19. Mermin ND (1965) Thermal properties of the inhomogeneous electron gas. Phys Rev 137:A1441–A1443
20. Staunton JB, Banerjee R, dos Santos Dias M, Deak A, Szunyogh L (2014) Fluctuating local moments, itinerant electrons, and the magnetocaloric effect: compositional hypersensitivity of FeRh. Phys Rev B 89:054427
21. Zemen J, Mendive-Tapia E, Gercsi Z, Banerjee R, Staunton JB, Sandeman KG (2017) Frustrated magnetism and caloric effects in Mn-based antiperovskite nitrides: Ab initio theory. Phys Rev B 95:184438
22. Mendive-Tapia E, Castán T (2015) Magnetocaloric and barocaloric responses in magnetovolumic systems. Phys Rev B 91:224421
23. Nikitin SA, Myalikgulyev G, Annaorazov MP, Tyurin AL, Myndyev RW, Akopyan SA (1992) Giant elastocaloric effect in FeRh alloy. Phys Lett A 171(3):234–236
24. Stern-Taulats E, Planes A, Lloveras P, Barrio M, Tamarit J-L, Pramanick S, Majumdar S, Frontera C, Mañosa L (2014) Barocaloric and magnetocaloric effects in $Fe_{49}Rh_{51}$. Phys Rev B 89:214105
25. Zhang XX, Tejada J, Xin Y, Sun GF, Wong KW, Bohigas X (1996) Magnetocaloric effect in $La_{0.67}Ca_{0.33}MnO_\delta$ and $La_{0.60}Y_{0.07}Ca_{0.33}MnO_\delta$ bulk materials. Appl Phys Lett 69(23):3596–3598
26. Caron L, Miao XF, Klaasse JCP, Gama S, Brück E (2013) Tuning the giant inverse magnetocaloric effect in $Mn_{2-x}Cr_xSb$ compounds. Appl Phys Lett 103(11):112404
27. Tegus O, Brück E, Zhang L, Dagula, Buschow KHJ, de Boer FR (2002) Magnetic-phase transitions and magnetocaloric effects. Phys B: Condens Matter 319(1):174 – 192

Chapter 5
Pair- and Four-Spin Interactions in the Heavy Rare Earth Elements

Rare earth materials are at the origin of many applications with increasing importance. However, their complicated magnetism arising from the strongly correlated f-electrons clearly sets an outstanding problem with significant interest on fundamental physics grounds that is not fully understood yet. This motivates the ab initio description of the diverse magnetism in the heavy rare earth (HRE) elements.

The lanthanides from gadolinium to lutetium order into an apparently complex array of magnetic phases [1] despite the common chemistry of their valence electronic structure ($5d^1 6s^2$ atomic configuration). Under ambient conditions they crystallise into hexagonal close packed structures and the number of localised f-electrons per atom increases from seven for Gd's half-filled shell through to Lu's complete set of fourteen, which causes the lanthanide contraction of the lattice [2]. The magnetism is complicated. When cooled through T_c, Gd's paramagnetic (PM) phase undergoes a second-order transition to a ferromagnetic (FM) state whereas, at T_N, Tb, Dy and Ho form incommensurate, helical antiferromagnetic (HAFM) phases where the magnetisation spirals around the crystal c axis. In panels (a) and (c) of Fig. 5.1 we illustrate these two magnetic phases, respectively. Moreover, when the temperature is lowered further both Tb and Dy undergo a first-order transition at T_t to a FM phase with basal plane orientation and Ho forms a conical HAFM ground state. Further exotic phases emerge when these metals are subjected to magnetic fields, which have been extensively studied in experiments [3–10]. For example, below its T_c Gd preserves its FM order as the strength of the magnetic field applied along its easy axis is increased. This is in sharp contrast to Tb [3, 11–13], Dy [4–7, 14, 15] and Ho [8, 10] above T_t, which first distort their HAFM order (dis-HAFM) before undergoing a first-order transition into a fan magnetic structure, shown in Fig. 5.1b, followed by a second-order transition to a FM state with further increase in the magnetic field. Contrary to a HAFM phase, in the fan structure the local moments swing back and forth around the magnetic field direction from one FM layer to another, similar to a pendulum in a mechanical metronome [1]. Dy and Ho also exhibit signs of additional spin-flip and vortex transitions associated with subtle changes in measured magnetisation curves [7, 8]. In this chapter we argue that much

© Springer Nature Switzerland AG 2020

E. Mendive Tapia, *Ab initio Theory of Magnetic Ordering*, Springer Theses,
https://doi.org/10.1007/978-3-030-37238-5_5

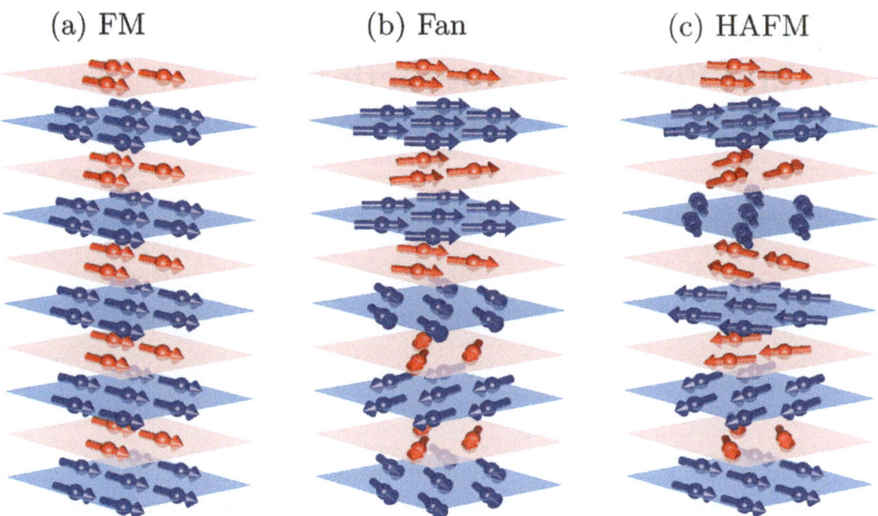

Fig. 5.1 **a** Ferromagnetic, **b** fan, and **c** helical antiferromagnetic magnetic phases observed in the heavy rare earth elements, illustrated by a ten- ferromagnetic layer scheme with single-site magnetisations perpendicular to the c axis. Colours blue and red are used to distinguish between the two non-equivalent atomic positions inside one crystallographic unit cell of the hexagonal close packed structure

of this diversity stems directly from the valence electronic structure that all the HRE elements share.

A simple classical spin model with pairwise exchange interactions, magnetic anisotropy contribution, and a Zeeman external magnetic field term, describes magnetic field-driven phase transitions in some antiferromagnetic insulators [16, 17]. If the local moments of a HAFM state are pinned by anisotropy and crystal field effects to spiral around a particular direction, the effect of a magnetic field causes a first-order transition to a fan or conical phase where the spins now oscillate about the field direction. In higher fields, the fan or cone angles smoothly decrease to zero in the FM state. Such a model applied to the HREs addresses only part of the phenomenon, however, since it fails to reproduce the first-order dis-HAFM to FM and second-order fan to FM transitions at low temperatures and fields. It misses a tricritical point in consequence. In the seminal work by Jensen and Mackintosh [18], where the formation of field-induced fan and helifan phases was investigated theoretically for the first time, the key aspects of the HRE magnetic phase diagrams were only reproduced if ad hoc temperature dependent pairwise exchange interactions were incorporated from a fit to spin wave measurements conducted at a series of temperatures. As shown in Chap. 3, our theory is designed to account for more complicated interactions than pairwise terms described by a classical Heisenberg Hamiltonian. From our study we find instead that much of the magnetic phase complexity is directly traced back to the behaviour of the valence electrons and their influence on multi-spin, or multi-site, effects.

Extensive experimental and theoretical investigations satisfactorily explain the onset of magnetic order from the PM state [19–26] via a detailed version of the famous Ruderman–Kittel–Kasuya–Yoshida (RKKY) pairwise interaction. The existence of nesting vectors $\mathbf{q}_{nest}$ separating parallel Fermi surface (FS) sheets of the valence electrons provokes a singularity in the conduction-electron susceptibility. This feature results in a $\mathbf{q}_{nest}$-modulated magnetic phase [1, 27], identified as a HAFM structure, incommensurate with the underlying lattice. The lanthanide contraction changes the FS topology [21, 25, 28] and acts as the decisive factor for the emergence of the nesting vectors. This has resulted in the construction of a universal crystallomagnetic phase diagram which links the magnetic ordering that emerges from the PM state to the specific c and a lattice parameters of a heavy lanthanide system [21, 23].

The prominence of RKKY interactions in the discussion of lanthanide magnetism promotes a deeper inspection of the common HRE valence electronic structure. As magnetic order among the local f-electron moments of the HREs develops with decreasing temperature and/or strengthening applied magnetic field the valence electron glue spin-polarises and qualitatively changes. This effect has a potentially profound feedback on the interactions between the magnetic moments and has wider relevance for other magnetic systems where the physics is also typically couched in RKKY terms, including giant magnetoresistive nanostructures [29], rare earth clusters [30], magnetic semiconductors [31] and spin glasses [32]. Evidently, it is precisely this response and feedback between magnetic ordering and the electronic structure that our theory is designed to calculate, as explained in Chaps. 3 and 4 (see in particular their respective Sects. 3.2 and 4.1). In the following we shall apply our SDFT-DLM approach, set out in Chaps. 2–4, with the aim to describe the effect from the heavy lanthanide valence electrons on the ordering of local magnetic moments and the subsequent creation of multi-spin interactions, showing how these determine the main features of the magnetic phase diagrams. Since these interactions are not imposed from the outset but derive from the theory naturally, they are tied to the FS topology which our ab initio model handles. In other words, the fact that the FS topology changes qualitatively with magnetic ordering demands the existence of non-negligible multi-site interactions. The main result from our study is the identification of a magnetic phase diagram reference, summarised in Fig. 5.2, against which the magnetic properties can be analysed to discriminate specific, subtle f-electron features.

Gd is a convenient prototype system owing to its seven localised f-electrons per atom in an S-state which form a large moment and the small crystal field and spin-orbit coupling effects that are prevalent. Hence, we choose Gd deliberately such that the common HRE valence electron effects on the magnetic properties are abstracted cleanly. Gd's relatively simple $L = 0$ f-electron configuration means that the main physics is unclouded by the actual details of the method used to deal with the strong f-electron correlations. For example, treating the occupied f-electrons as part of the core with LDA+U [33, 34] or with the self-interaction correction (SIC) [35, 36] should lead to the same generic phase diagram features shown in Fig. 5.2. Selecting the c and a lattice constants for other elements makes the analysis appropriate to Tb,

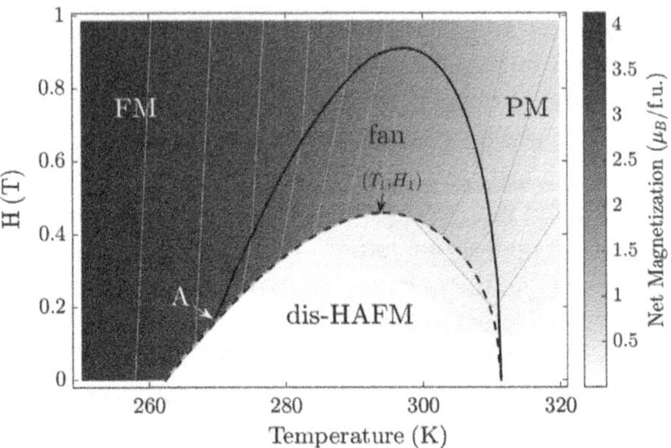

Fig. 5.2 The generic magnetic T-H phase diagram for a heavy lanthanide metal for **H** applied along the easy direction constructed from theory. Continuous (discontinuous) lines correspond to second- (first-) order phase transitions and a tricritical point is marked ('A'). The calculations were performed for the prototype Gd with lattice constants appropriate to Dy

Dy, Ho etc. In particular, Fig. 5.2 shows the results for Gd using the lattice parameters appropriate to Dy [23].

The theory has the advantage that the valence electronic structure can be monitored as a function of local moment disorder [37, 38]. This is highly pertinent owing to the recent development of advanced time-dependent spectroscopy techniques. Time-resolved resonant X-ray and ultrafast magneto-optical Kerr studies confirm the central tenet of our theory that the dynamics of the HRE core-like f-electrons are on a much longer time-scale than the excitations of the valence electrons [39]. Time- and angle-resolved photoemission (ARPES) studies have demonstrated the differing dynamics of spin-polarised valence states in correlated materials [40–42]. Figure 5.3 shows the Bloch spectral function at the FS of Gd with Dy's lattice attributes within the DLM picture. This object is directly obtained from the Green's function and by lattice Fourier transforming Eq. (2.72), given in Chap. 2, as [37, 38, 43, 44],

$$A(\mathbf{k}, E) = -\frac{1}{\pi} \mathrm{ImTr} \sum_{nn'} \exp\left[i\mathbf{k} \cdot (\mathbf{x}_n - \mathbf{x}_{n'})\right] \int \mathrm{d}\mathbf{r} G^+(\mathbf{r}, \mathbf{r}, E). \qquad (5.1)$$

Here the indices n and n' visit all magnetic positions in the lattice, i.e. $\mathbf{x}_n \equiv \mathbf{R}_t + \mathbf{r}_s$ in the notation introduced in Sect. 3.4, and the dependence of the Green's function on these position vectors is given in Eq. (2.67). Figure 5.3a shows the FS when the moments are randomly oriented in the paramagnetic state. Note that the broadening is caused by the spin fluctuation disorder. The nesting vectors responsible for the onset of Dy's HAFM state below T_N are clearly seen [23]. Figures 5.3b, c show the FS where now the local f-electron moments are oriented on average to produce an overall net average magnetisation of 54% of the $T = 0\,\mathrm{K}$ saturation value. The FS is spin-polarised and neither majority nor minority spin component continues to show

(a)   **(b)**   **(c)**

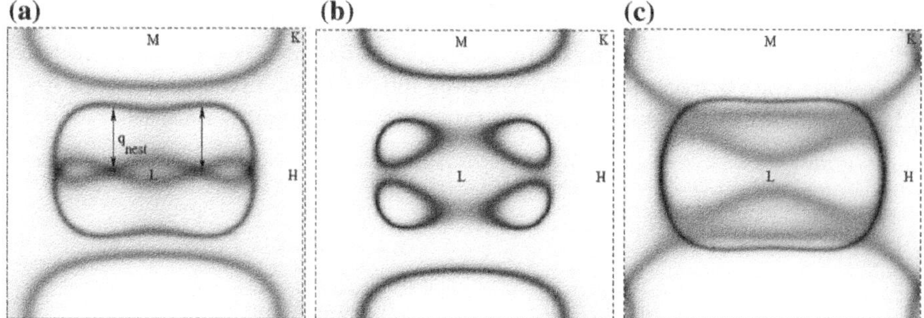

Fig. 5.3 The Bloch spectral function $A(\mathbf{k}, E)$ in the $LMHK$ plane at the Fermi energy for Gd with Dy's lattice attributes for **a** the PM state and resolved into **b** majority spin and **c** minority spin components when there is an overall net average magnetisation of 54% ($m_{\mathrm{FM}} = 0.54$) of the $T = 0\,\mathrm{K}$ saturation value in the FM state. This is the value in our calculations (Fig. 5.2) in the FM phase just below the temperature T_t of the HAFM-FM first-order transition. $\mathbf{q}_{\mathrm{nest}}$ indicates the nesting wave-vector of the FS of the PM state and the shading represents the broadening from thermally induced local moment disorder

nesting. This dramatic change of FS topology hints at the valence electron's role in the HAFM-FM metamagnetic transition. It also concurs with conclusions drawn from Döbrich et al. [24] angle-resolved photoemission measurements pointing to the magnetic exchange splitting of the FS as the principal mechanism for the fading of the nesting vectors [27]. This causes the stability of the FM phase at the ground state in Tb and Dy.

To follow the repercussions of this insight we continue by specifying our generalised Grand Potential $\tilde{\Omega}(\{\hat{e}_n\})$ and studying the effect of magnetic ordering from the perspective of magnetic interactions. Of course, the way to proceed is to average this quantity over many magnetic configurations prescribed by the probabilities $\{P_n(\hat{e}_n)\}$, i.e. choices of $\{\mathbf{m}_n\}$, that adequately describe the FM, HAFM, fan and in-between magnetic structures in the HREs. This is the method shown in Chaps. 3 and 4. Note that due to the particularities of these magnetic phases, changing the index n to explore different atoms within one ferromagnetic layer gives equivalent magnetic local order parameters. Firstly, we carried out SDFT-DLM calculations for Gd for the c and a lattice parameters appropriate to Gd, Tb, Dy and Ho and in each case calculated charge and magnetisation densities self-consistently for the PM state ($\{\mathbf{m}_n = 0\}$). Since the local moments arise from the fairly well localised f-electrons the RSA can be invoked so that the PM potentials have been used at different magnetic orderings, owing to their good local moment nature (see Sect. 3.3.2). The SIC was used to capture the strong correlations of the f-electrons [23, 35, 37] and a local moment of size $\mu \approx 7.3\mu_B$ established on each Gd site as a result of this calculation. We then divided the hexagonal lattice into 10 layer stacks and specified sets of $\{\mathbf{m}_n\}$ ($n = 1, \ldots, 10$) values in order to describe the different magnetic orderings. For each c and a pair and set of $\{\mathbf{m}_n\}$ we calculated the $\mathbf{h}_n^{\mathrm{int}} = -\nabla_{\mathbf{m}_n} \langle \Omega^{\mathrm{int}} \rangle_0$ values using the effective one electron PM potentials and thoroughly followed the methodology described in Chap. 4 to calculate pairwise and higher exchange constants. Note that owing to the nature of these magnetic phases, $\langle \Omega^{\text{int}} \rangle_0$ depends on 10 vectors $\{\mathbf{m}_n\}$ that describe the 10 layer stacks. By repeating the calculation for many sets of $\{\mathbf{m}_n\}$ and careful analysis we found $\langle \Omega^{\text{int}} \rangle_0$ to fit very well the expression

$$
\begin{aligned}
\langle \Omega^{\text{int}} \rangle_0 &= -\sum_{n,n'} \left(\mathcal{J}_{nn'} + \sum_{n'',n'''} \mathcal{K}_{nn',n''n'''} \mathbf{m}_{n''} \cdot \mathbf{m}_{n'''} \right) \mathbf{m}_n \cdot \mathbf{m}_{n'} \\
&= -\sum_{n=1}^{10} \Bigg\{ \left(J_1 m_n^2 + K_1 m_n^4 \right) - \sum_{n'=1}^{5} \Big[\frac{1}{2} J_{1+n'} \mathbf{m}_n \cdot (\mathbf{m}_{n+n'} + \mathbf{m}_{n-n'}) \\
&\quad + K_{n'+1}(\mathbf{m}_n \cdot \mathbf{m}_{n+n'} + \mathbf{m}_n \cdot \mathbf{m}_{n-n'})(m_n^2 + m_{n+n'}^2 + m_{n-n'}^2) \Big] \\
&\quad - \frac{1}{4} K' \Big[(\mathbf{m}_n \cdot \mathbf{m}_{n+1})(\mathbf{m}_{n+2} \cdot \mathbf{m}_{n+3} + \mathbf{m}_{n-1} \cdot \mathbf{m}_{n-2}) \\
&\quad + (\mathbf{m}_n \cdot \mathbf{m}_{n-1})(\mathbf{m}_{n-2} \cdot \mathbf{m}_{n-3} + \mathbf{m}_{n+1} \cdot \mathbf{m}_{n+2}) \Big] \Bigg\}.
\end{aligned}
\tag{5.2}
$$

In other words, we tested that $\{\mathbf{h}_n^{\text{int}}\}$ is well described by this expression and hence extracted the effect of pairwise and quartic interactions, $\mathcal{J}_{nn'}$ and $\mathcal{K}_{nn',n''n'''}$, at a very satisfactory numerical level. We also checked that higher order terms were vanishingly small. We emphasise that these coefficients connect all magnetic sites from one FM layer to another, labelled by the sub-indices, i.e. they compactly contain the different individual local moment interactions between layers. The coefficients $\mathcal{J}_{nn'}$'s and $\mathcal{K}_{nn',n''n'''}$, structured by the six constants $\{J_n, n = 1, \ldots, 6\}$ and the seven constants $\{K_n, K'', n = 1, \ldots, 6\}$, respectively, arise from the mutual feedback between local moment magnetic order and the spin polarised valence electrons illustrated in Fig. 5.3. Specifically, we obtained them by exploring the values $\lambda_n = \beta h_n = 0, 0.1, 0.5, 1.0, 1.5, 2.0, 3.0, 5.0, 7.0,$ and 10, for each magnetic structure of interest, i.e. for FM, HAFM, helifan and fan orderings. This corresponds to the study of order parameters with magnitudes ranging from $m_n = 0$ to 0.9. We also sampled additional arrangements where only two layers have non-zero λ_n values separated by several distances and where the angles of the magnetisation directions between the two layers are also varied. We have found this to be a crucial step in order to elucidate the correct form of Eq. (5.2). As a result, the number of magnetic configurations used for each lattice spacing was 95. The data used to fit the 17 constants was totally composed, therefore, by the corresponding internal fields $\{\mathbf{h}_n^{\text{int}} = \nabla_{\mathbf{m}_n} \langle \Omega^{\text{int}} \rangle_0, n = 1, \ldots, 10\}$ prescribed by the 10 vectors $\{\mathbf{m}_n, n = 1, \ldots, 10\}$ and calculated at every site for each of these 95 configurations. In Table 5.1 we give the values obtained.

For a phase diagram such as Fig. 5.2 we construct the Gibbs free energy from Eq. (3.26) and add a uniaxial anisotropy term $F_u(\mathbf{m}_n) = F_0 \langle (\hat{e}_{n,i} \cdot \hat{z})^2 \rangle$ [45] with strength F_0 [46] with the solely intention to fix the easy plane magnetisation, that is

Table 5.1 The table shows the pair ($\{J_n\}$) and quartic ($\{K_n\}$, K') constants in meV/f.u. units obtained from the fitting of the SDFT-DLM data of Gd with the lattice attributes of Gd, Tb, Dy, and Ho. These have been scaled by the de Gennes factor. The lattice parameters used in the calculation in atomic units are also shown

	Gadolinium	Terbium	Dysprosium	Holmium
a	6.613	6.605	6.587	6.565
c	10.78	10.44	10.35	10.31
c/a	1.630	1.580	1.572	1.570
J_1	20.73	17.28	16.70	15.96
J_2	21.94	22.63	22.29	21.68
J_3	5.576	0.572	4.172	4.222
J_4	-2.244	-3.275	-3.589	-3.728
J_5	0.466	-0.348	-0.559	-0.507
J_6	-1.729	-2.736	-2.846	-3.083
K_1	0.302	0.534	0.544	0.545
K_2	-0.162	-0.427	-0.411	-0.385
K_3	-0.003	0.088	0.123	0.124
K_4	0.303	0.337	0.333	0.339
K_5	-0.187	-0.095	-0.094	-0.092
K_6	0.154	0.177	0.179	0.184
K'	0.177	1.020	1.118	1.231

$$\mathcal{G}_1 = \langle \Omega^{\text{int}} \rangle_0 - \sum_n [\mu \mathbf{m}_n \cdot \mathbf{H} + T S_n(\mathbf{m}_n) + F_u(\mathbf{m}_n)]. \tag{5.3}$$

An analytical expression for the uniaxial anisotropy can be derived by performing the integral $\langle (\hat{e}_n \cdot \hat{z})^2 \rangle = \int P_n(\hat{e}_n)(\hat{e}_n \cdot \hat{z})^2 \mathrm{d}\hat{e}_n$, which gives

$$F_u(\mathbf{m}_n) = F_0 \left[(\hat{\boldsymbol{\lambda}}_n \cdot \hat{z})^2 - \frac{1}{\lambda_n} \left(\frac{-1}{\lambda_n} + \coth \lambda_n \right) \left(3(\hat{\boldsymbol{\lambda}}_n \cdot \hat{z})^2 - 1 \right) \right]. \tag{5.4}$$

By comparing Gibbs free energies of the FM, HAFM, conical, and fan structures we constructed the T-H phase diagrams. We examined the stability of helifan phases too but they were not found to globally minimise $\mathcal{G}_1$ at any temperature and strength of $\mathbf{H}$. Figure 5.2 summarises the results for Gd with the lattice spacing appropriate to Dy when $\mathbf{H}$ was applied along the easy direction. Continuous/dashed lines correspond to second-/first- order phase transitions. The single-site uniaxial anisotropy imposed has a typical magnitude of $F_0 = +6.3$ meV/site [46] observed in experiment, which precluded the conical phase when the magnetic field was applied in the easy ab plane. The figure reproduces all the main features that the experimentally measured magnetic phase diagrams of heavy lanthanide metals and their alloys have in common. There is the first-order HAFM-FM transition in the absence of $\mathbf{H}$ at T_t. Then for increasing values of **H** applied along the easy direction the helical structure initially distorts before transforming to the fan structure. Increasing **H** further stabilises the FM phase. There is, also in line with experimental findings, a second-order transition from the fan to FM phase in finite field on cooling and we find a tricritical point which is marked (A) in Fig. 5.2. At a value $m_{\mathrm{FM}} = 0.503$ for all local order parameters $\{m_n\}$, i.e. describing the FM state, the system undergoes a first-order transition from a HAFM to FM state in zero field at $T_t = 262\,\mathrm{K}$, which correlates with the FS topological changes depicted in Fig. 5.3. This is produced by the presence of the four-site $\mathcal{K}_{nn',n''n'''}$'s constants. For example, when they are set to zero $\mathcal{K}_{nn',n''n'''} = 0$, and so the feedback between the valence electronic structure and lanthanide local f-electron magnetic moment ordering is omitted, the calculated phase diagram is radically altered and qualitatively at odds with experimental results [7], as shown in Fig. 5.4. A direct consequence of this is that the FM phase and the tricritical point (A) do not appear at low temperatures and fields anymore. As an important remark, we point out that the phase diagram features are largely insensitive to the magnitude of the anisotropy term, leading to the same results when the value of F_0 is changed over a large range including the experimental values reported in the literature [46]. For example, the difference between the PM-HAFM and HAFM-FM transition temperatures, $T_N - T_t$, changes by less than $1\,\mathrm{K}$ when F_0 is varied from 3 to $15\,\mathrm{meV/site}$. Similarly, the value of the critical field H_1 indicated in Fig. 5.2 changes by less than $2\,\mathrm{mT}$ for the same range of values.[1]

For completeness, in Fig. 5.5 we show the magnetic phase diagram when the magnetic field is applied along the hard axis. The figure illustrates how the cone structure, which consists in a HAFM phase with a perpendicular ferromagnetic component [1], stabilises as the magnetic field is increased. Interestingly, the transition from the conical HAFM to the FM/PM state changes from second- to first-order at higher and lower temperatures, respectively, originating another tricritical point.

In terms of lattice Fourier transforms (summing over all magnetic positions) Eq. (5.2) can be written as

$$
\langle \Omega^{\mathrm{int}} \rangle_0 = - N \sum_{\mathbf{q}} \left(\mathcal{J}(\mathbf{q}) + \sum_{\mathbf{q}'} \mathcal{K}(\mathbf{q}, \mathbf{q}') \left(\mathbf{m}(\mathbf{q}') \cdot \mathbf{m}(-\mathbf{q}') \right) \right) \left(\mathbf{m}(\mathbf{q}) \cdot \mathbf{m}(-\mathbf{q}) \right)
$$

$$
= - \sum_{\mathbf{q}, nn'} \left(\mathcal{J}_{nn'} + \sum_{\mathbf{q}', n''n'''} \mathcal{K}_{nn',n''n'''} e^{-i\mathbf{q}' \cdot (\mathbf{x}_{n''} - \mathbf{x}_{n'''})} \left(\mathbf{m}(\mathbf{q}') \cdot \mathbf{m}(-\mathbf{q}') \right) \right)
$$

$$
\times\, e^{-i\mathbf{q} \cdot (\mathbf{x}_n - \mathbf{x}_{n'})} \left(\mathbf{m}(\mathbf{q}) \cdot \mathbf{m}(-\mathbf{q}) \right),
$$

$$
(5.5)
$$

where N is the number of magnetic atoms in the system. Equation (5.5) directly defines effective pair interactions affecting and mediated by the valence electrons in the presence of long-range magnetic order $\mathbf{m}(\mathbf{q}')$, i.e. showing mode-mode coupling,

---

[1]In addition, we checked that using either a fully relativistic or scalar relativistic version of our SDFT-DLM method produces the same f-electron magnetic ordering—valence electronic structure linkage.

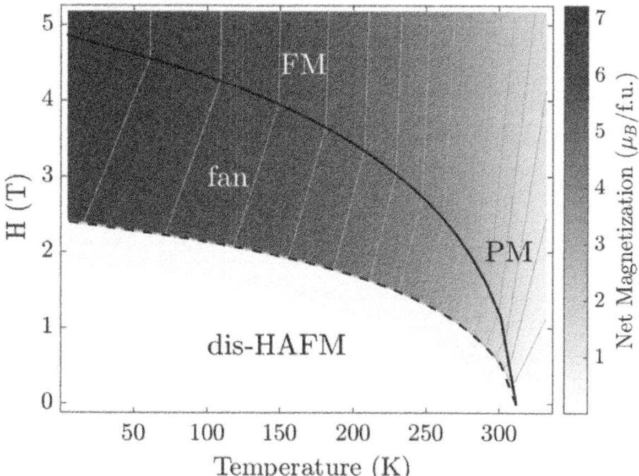

Fig. 5.4 The phase diagram of Gd with the lattice spacing of Dy when the quartic coefficients are set to zero and the magnetic field is applied along the easy direction. Continuous (discontinuous) lines correspond to second- (first-) order transitions

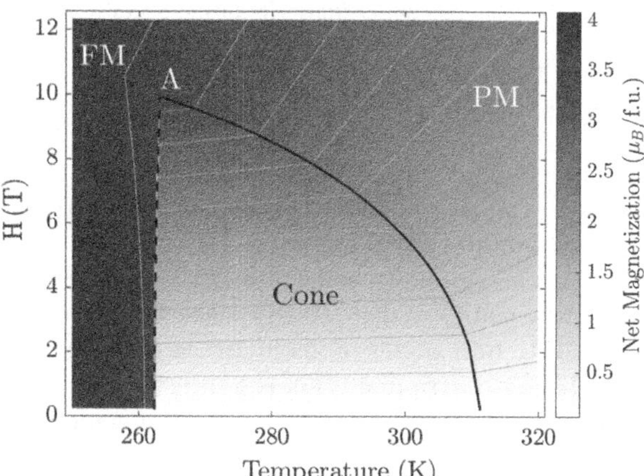

Fig. 5.5 The phase diagram of Gd with the lattice spacing of Dy when the magnetic field is applied along the hard direction. Continuous (discontinuous) lines correspond to second- (first-) order transitions and a tricritical point is marked (A)

$$\mathcal{J}^{\text{eff}}_{\mathbf{q}',nn'} = \mathcal{J}_{nn'} + \sum_{n''n'''} \left(\mathcal{K}_{nn',n''n'''} e^{-i\mathbf{q}'\cdot(\mathbf{x}_{n''}-\mathbf{x}_{n'''})} \left(\mathbf{m}(\mathbf{q}') \cdot \mathbf{m}(-\mathbf{q}') \right) \right). \tag{5.6}$$

When $\mathbf{q}' = \mathbf{0}$ it follows that $\mathbf{m}(\mathbf{0}) = \mathbf{m}_{\text{FM}}$ so that the equation above becomes

$$\mathcal{J}^{\text{eff}}_{nn'} = \mathcal{J}_{nn'} + \sum_{n''n'''} \mathcal{K}_{nn',n''n'''} m^2_{\text{FM}}, \tag{5.7}$$

which describes the influence of the four-site terms when ferromagnetic ordering, quantified by m_{FM}, develops. In the terminology of the elucidated constants $\{J_n\}$, $\{K_n\}$ and K', from Eq. (5.7) the effective quadratic coefficients are

$$\begin{aligned}
J^{\text{eff}}_1 &= J_1 + K_1 m^2_{\text{FM}} \\
J^{\text{eff}}_{n+1} &= J_{n+1} + \left(6K_{n+1} + 2K'\right) m^2_{\text{FM}} \qquad \text{for } \{n = 1, \dots, 5\}
\end{aligned} \tag{5.8}$$

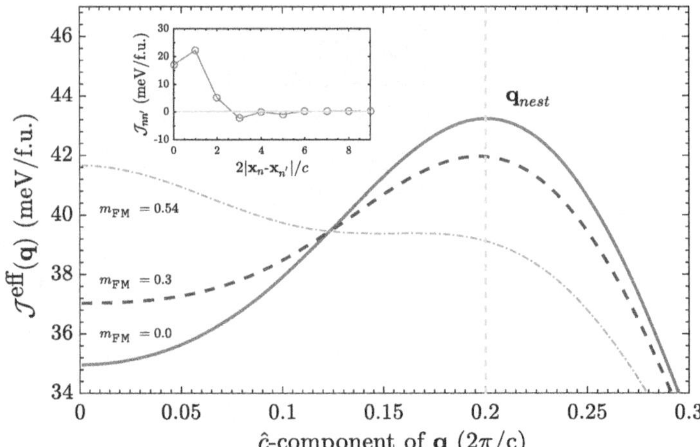

Fig. 5.6 The lattice Fourier transform of the effective pair interactions $\mathcal{J}^{\text{eff}}(\mathbf{q})$ (red line) when $m_{\text{FM}} = 0$ and its change when the FS is spin polarised, for finite m_{FM} (dashed blue line for $m_{\text{FM}} = 0.3$ and dot-dashed green line for $m_{\text{FM}} = 0.54$). The inset shows the dependence of the pair interactions $\mathcal{J}_{nn'}$ on separation $x_{nn'} = |\mathbf{x}_n - \mathbf{x}_{n'}|$ for $m_{\text{FM}}=0$

Figure 5.6 shows the lattice Fourier transform of Eq. (5.7) for different values of m_{FM} relevant to Figs. 5.2 and 5.3, revealing the effect of the valence electron spin polarisation.[2] As shown in the inset, for $m_{\text{FM}} = 0$ the interactions have a long-ranged oscillatory nature so that $\mathcal{J}^{\text{eff}}(\mathbf{q})$ peaks at $\mathbf{q}_{\text{nest}} \approx 0.2\frac{2\pi}{c}\hat{c}$ (full red line), which corresponds to turn angles of about $36°$ in good agreement with experiment [24] and our 10-layer description. This is a direct consequence of the FS nesting shown in Fig. 5.3a and which drives the HAFM magnetic order.[3] We also show $\mathcal{J}^{\text{eff}}(\mathbf{q})$ for non-zero m_{FM}. When $m_{\text{FM}}=0.54$ (green, dot-dashed lines), the value in the FM phase just below T_t, $\mathcal{J}^{\text{eff}}(\mathbf{q})$ peaks at $\mathbf{q} = \mathbf{0}$ showing how the development of long-range magnetic order has favoured the shift towards ferromagnetism. This confirms the role of the spin-polarised valence electrons and altered FS topology exemplified in Fig. 5.3b, c.

We can adapt this Gd-prototype model to a specific heavy lanthanide element or alloy by using suitable lanthanide-contracted lattice constant values [23] and accounting for the specific f-electron configuration. The Gd ion has orbital angular momentum $L = 0$ and negligible spin orbit coupling effects. LS-coupling, however, is important for the HREs. A simple measure to account for the different total angular momentum values, J, is to scale the $\mathcal{J}_{nn'}$ and $\mathcal{K}_{nn',n''n'''}$ interactions with the famous

---

[2]Note that contrary to the lattice Fourier definition given in Eq. (3.68), here we sum over all magnetic positions, i.e. $\mathcal{J}^{\text{eff}}(\mathbf{q}) = \sum_{nn'} \mathcal{J}^{\text{eff}}_{nn'} \exp\left[\mathbf{q} \cdot (\mathbf{x}_n - \mathbf{x}_{n'})\right]$.

[3]As an important note, we verified that the paramagnetic state is unstable to the formation of a HAFM structure prescribed by a 10 layer periodicity from the paramagnetic limit analysis based on the calculation of the direct correlation function, as described in Sect. 3.4. The long-range constants ($n - n' > 6$) shown in the inset of Fig. 5.6 have been obtained from this calculation.

Table 5.2 Application of the theory to Tb, Dy and Ho. The values of T_N, T_t and the T of the highest **H** of the dis-HAFM phase (T_1 and H_1 in Fig. 5.2) are compared to experiment. Theoretical estimate of the tricritical point (A) is also given. Gd has a PM-to-FM second-order transition at $T_C = 274$ K ($T_C = 293$ K in experiment [1]). We remark that the HRE metals Er and Tm, which have larger lanthanide contractions, form incommensurate HAFM structures at low T and show no transitions to FM states [1], in agreement with the trends predicted here

Element	Tb	Dy	Ho
references	[3, 11–13]	[4, 7, 14]	[8, 47, 48]
T_N(K)	214	145	94
T_N^{exp}(K)	229	180	133
T_t(K)	206	90	–
T_t^{exp}(K)	222	90	20
T_1(K)	211	129	65
T_1^{exp}(K)	224–226	165–172	110
H_1(T)	0.03	0.43	1.03
H_1^{exp}(T)	0.02–0.03	1.1–1.2	3.0
T_A(K)	207	94	–
H_A(T)	0.01	0.07	–

de Gennes factor $(g_J - 1)2J(J + 1)$, where g_J is the Landé g factor [1]. Note that the quartic constants should be scaled by the square of $(g_J - 1)2J(J + 1)$ [1]. We applied this treatment to Gd, Tb, Dy and Ho and each metal except Gd showed the generic features of the form of Fig. 5.2 consistent with experiment, as can be seen in Fig. 5.7. Indeed, Gd's phase diagram, shown in panel (a), presents a single second-order transition between PM and FM phases with a Curie temperature of T_C=274K, in agreement with experiment [1].

Finally, in Table 5.2 we compare our results with those available from experiment for T_N and T_t, and the values of T for the highest **H** in the stabilisation of the dis-HAFM structure. We also give a theoretical estimate of the tricritical point. The comparison overall shows that the theory correctly captures trends and transition temperature and field magnitudes. When the c and a values are further decreased our model predicts that the FM phase does not appear at low T and fields in agreement with some experiments [49]. Discrepancies between the model and experiment can further highlight where f-electron correlation [50] effects are leading to more complicated physics. For example, the complex spin slip phases in Ho reported at low temperatures [8, 10] are not found in our model. The same applies to the vortex and helifan phases inferred from some experimental studies [7, 18]. It is worth mentioning that the results for Ho do not exhibit the both low temperature FM phase and the tricritical point. It seems that the model overestimates the effect of the lanthanide contraction in this element. As a final note, although Evenson and Liu [20] maintained that the first-order HAFM-FM transition is driven by a magnetoelastic effect, here we show rather that while there is a magnetostructural coupling it is not necessary for the transition.

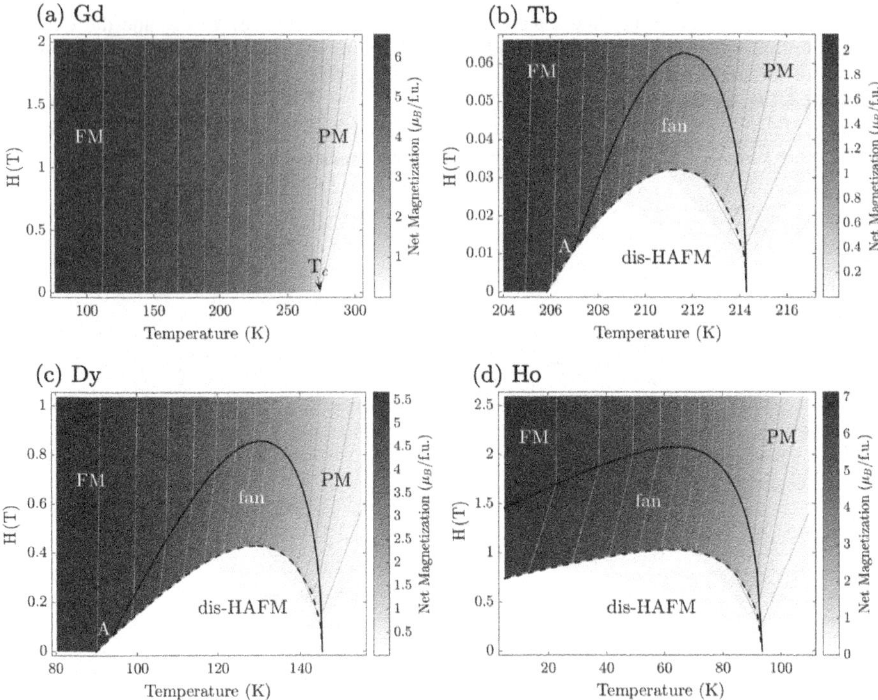

Fig. 5.7 The magnetic phase diagram constructed for Gd with the lattice attributes of **a** Gd, **b** Tb, **c** Dy, and **d** Ho. The de Gennes factor has been used to scale the quadratic and quartic coefficients (for the later we have used the square of the de Gennes factor) and the magnetic field is applied along the easy direction. Continuous (discontinuous) lines correspond to second- (first-) order phase transitions and a tricritical point (A) is marked when it exists

5.1 The Magnetocaloric Effect in the Heavy Rare Earth Elements

In this section we apply the theory presented in Chap. 4, Sect. 4.4, to calculate caloric effects in the HRE elements. We build on our Gd-prototype model to describe different heavy lanthanide elements using the appropriate de Gennes factor to account for the effect of different total angular momentum values on the magnetic interactions. In particular, we study the MCE produced by the application of an external magnetic field **H** along the easy plane fixed by the magnetocrystalline anisotropy, and compare it with experiment.

Whilst the prediction of caloric responses in Gd only requires the description of the single second-order PM-FM phase transition, the heavier elements Tb, Dy, and Ho set up a challenging scenario owing to their more exotic and complicated magnetism. Additional first-order transitions to a HAFM phase and subsequent tricritical points, as established in Fig. 5.2 and observed in experiment [3, 7], are at the centre of the

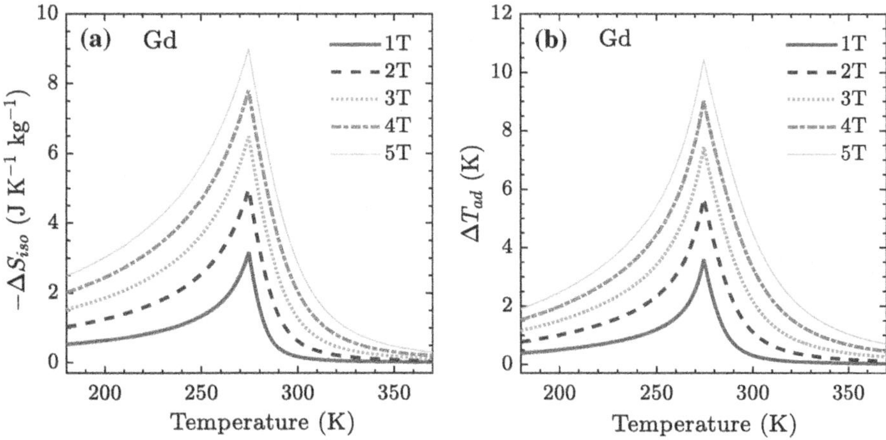

Fig. 5.8 The MCE quantified by **a** the isothermal entropy change and **b** the adiabatic temperature change of Gd with the lattice attributes of Gd. The figure shows the results in the vicinity of the second-order PM-FM phase transition for increasing values of the applied magnetic field

study. For example, by inspecting the behaviour of experimentally measured isofield magnetisation curves in Dy (see Fig. 4 of Ref. [7]) one can observe that the gradient of the magnetisation with respect to T changes sign in a non simple manner. By recalling Eq. (4.23), X being the magnetisation and Y the magnetic field, it follows that this situation is clearly susceptible to generate both conventional and inverse caloric effects, i.e. negative and positive isothermal entropy changes, associated with different regions of the magnetic phase diagrams. The MCE and its dependence on the magnetic field are, therefore, expected to be complicated.

To evaluate the MCE it is necessary to obtain the total entropy as introduced in Eq. (4.30). Firstly, we proceed by calculating the magnetic term $S_{\mathrm{mag}}(T, \mathbf{H})$ from Eqs. (3.23) and (3.24) in Chap. 3 at different values of T and $\mathbf{H}$. Note that this calculation is readily given by the values of $\{\mathbf{m}_n\}$ at each point $(T, \mathbf{H})$ of the magnetic phase diagram. Similarly, S_{elec} is obtained by providing the dependence of the density of states at the Fermi energy $n(E_\mathrm{F})$ on $\{\mathbf{m}_n\}$ into Eq. (4.25), which is directly available from SDFT-DLM machinery. Moreover, we calculate the vibrational contribution S_{vib} by applying the Debye model introduced in Eq. (4.27) and using a Debye temperature θ_D of 180 K for the HRE elements [51, 52]. We remark that experimentally it has been established that the dependence of θ_D on the magnetic ordering is relatively small [51, 53]. We tested that the results are effectively insensitive to the variation of θ_D measured at different temperatures and lanthanides, from Gd to Ho.

After performing relevant calculations we have found that in general the changes of $n(E_\mathrm{F})$ at different points in the phase diagrams are of the order of $10\,\mathrm{eV}^{-1}$/atom as the largest. These values lead to electronic entropy changes being approximately $\Delta S_{elec} \approx 0.3\,\mathrm{J\,K}^{-1}\mathrm{kg}^{-1}$ or below, which is about an order of magnitude lower than the ones originated from magnetic orientational sources. Figures 5.8 and 5.9 show the results when ignoring S_{elec} for Gd with the lattice attributes and the appro-

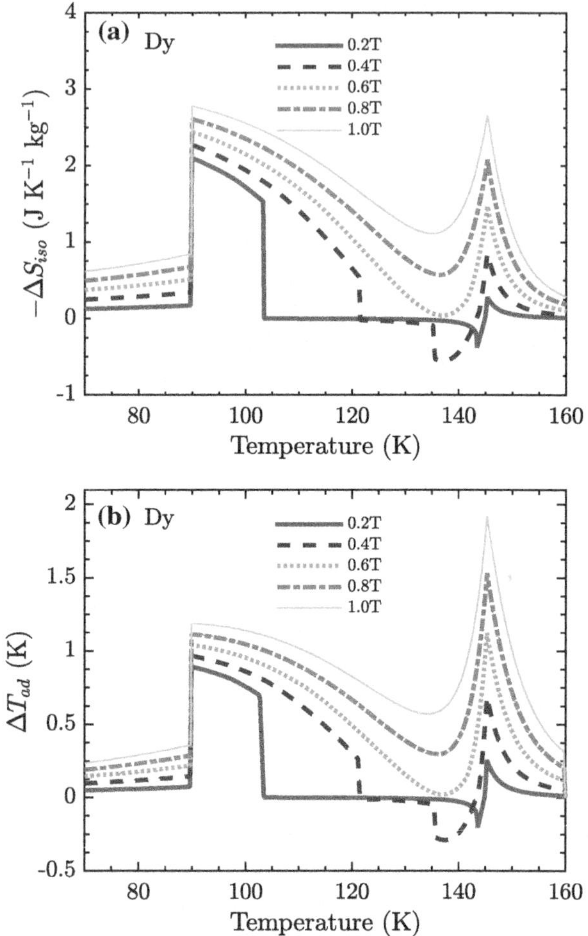

Fig. 5.9 The MCE quantified by **a** the isothermal entropy change and **b** the adiabatic temperature change of Gd with the lattice attributes of Dy. The figure shows the results in a temperature range comprising the PM-HAFM and HAFM-FM phase transitions for increasing values of the applied magnetic field. The appropriate de Gennes factor for Dy scaling the pair- and quartic constants has been applied

priate de Gennes scaling of Gd and Dy, respectively. The plots for Tb and Ho present qualitatively same traits compared with Dy but following the features of their respective phase diagrams. Both figures show an excellent qualitative agreement with experiment and a very good level in quantitative comparison [7, 54, 55]. We especially emphasise the correct description of the complicated MCE behaviour of Dy in sharp contrast with the simplicity of Gd. Clearly, the different features in the diagrams between Gd and Dy largely change their magnetocaloric responses. Our calculations correctly predict the regions where the effect changes from conventional ($-\Delta S_{iso} > 0$) to inverse ($-\Delta S_{iso} < 0$), as can be seen by comparing panel (a) in Fig. 5.9 with Fig. 22 in Ref. [7]. We conclude that the most significant features of the MCE in the HRE elements are described from the particularities of their magnetic phase diagrams, which in turn are fundamentally linked to the valence electrons and the consequent pair- and quartic interactions.

5.2 Summary and Conclusions

In this chapter we have used our SDFT-DLM theory, presented in Chaps. 2–4, to study the diverse magnetism in the HRE elements. The method applied is based on the calculation of internal magnetic fields emerging when specific magnetic ordering is imposed on the electronic structure. In particular, we have explored ferromagnetic and long wave modulated helical antiferromagnetic phases, as well as the effect of an external magnetic field, giving rise to distorted helical antiferromagnetic and fan structures. We claim a dominant role for the valence electrons in the constructed temperature-field magnetic phase diagrams of the heavy rare earth metals. Our ab initio theory incorporates lattice structural effects coming from the lanthanide contraction on this glue and makes the link between the changing topology of the Fermi surface, observed experimentally, and the evolving long range magnetic order of the f-electron moments, which triggers the first-order transition between helical antiferromagnetic and ferromagnetic states. Tricritical points are also predicted. This generic valence electron effect produces pairwise and four-site constants describing magnetic-order-dependent interactions among the localised f-electron moments and rules out the necessity to invoke ad hoc temperature dependent effective interactions or magnetostrictive effects. A simple de Gennes factor scaling of the interactions along with a phenomenological measure of magnetocrystalline anisotropy to fix the easy magnetisation plane enables this model to be applied broadly to the heavy rare earths. We propose the model as a filter to identify subtle lanthanide f-electron correlation effects for further scrutiny. To finalise the chapter we also calculated the magnetocaloric effect in Gd and Dy, obtaining an excellent agreement with experiment. We concluded that the caloric responses are mainly linked to the specific traits in the respective magnetic phase diagrams, which at the same time are fundamentally described by the pairwise and four-site interactions.

References

1. Jensen J, Mackintosh AR (1991) Rare earth magnetism, estructure and excitations. Clarendon, Oxford
2. Gschneidner KA, Eyring LR (1978) Handbook on the physics and chemistry of rare Earths. North-Holland, Amsterdam
3. Zverev VI, Tishin AM, Chernyshov AS, Ya Mudryk KA, Jr G, Pecharsky VK (2014) Magnetic and magnetothermal properties and the magnetic phase diagram of high purity single crystalline terbium along the easy magnetization direction. J Phys. Condens Matter 26(6):066001
4. Herz R, Kronmüller H (1978) Field-induced magnetic phase transitions in dysprosium. J Magn Magn Mater 9(1):273–275
5. Alkhafaji MT, Ali N (1997) Magnetic phase diagram of dysprosium. J Alloys Compd 250(1):659–661
6. Kida Y, Tajima K, Shinoda Y, Hayashi K, Ohsumi H (1999) Effect of magnetic field on crystal lattice in dysprosium studied by X-ray diffraction. J Phys Soc Jpn 68(2):650–654

7. Chernyshov AS, Tsokol AO, Tishin AM, Gschneidner KA, Pecharsky VK (2005) Magnetic and magnetocaloric properties and the magnetic phase diagram of single-crystal dysprosium. Phys Rev B 71:184410

8. Zverev VI, Tishin AM, Min Z, Mudryk Y, Gschneidner KA Jr, Pecharsky VK (2015) Magnetic and magnetothermal properties, and the magnetic phase diagram of single-crystal holmium along the easy magnetization direction. J Phys: Conden Matter 27(14):146002

9. Cowley RA, Bates S (1988) The magnetic structure of holmium. I. J Phys C: Solid State Phys 21(22):4113

10. Ali N, Willis F, Steinitz MO, Kahrizi M, Tindall DA (1989) Observation of transitions to spin-slip structures in single-crystal holmium. Phys Rev B 40:11414–11416

11. Jiles DC, Palmer SB, Jones DW, Farrant SP, Gschneidner KA Jr (1984) Magnetoelastic properties of high-purity single-crystal terbium. J Phys F: Metal Phys 14(12):3061

12. Jiles DC, Blackie GN, Palmer SB (1981) Magnetoelastic effects in terbium. J Magn Magn Mater 24(1):75–80

13. Kataev GI, Sattarov MR, Tishin AM (1989) Influence of commensurability effects on the magnetic phase diagram of terbium single crystals. Phys Status Solidi (a) 114(1):K79–K82

14. Yu J, LeClair PR, Mankey GJ, Robertson JL, Crow ML, Tian W (2015) Exploring the magnetic phase diagram of dysprosium with neutron diffraction. Phys Rev B 91:014404

15. Chernyshov AS, Mudryk Y, Pecharsky VK, Gschneidner KA (2008) Temperature and magnetic field-dependent x-ray powder diffraction study of dysprosium. Phys Rev B 77:094132

16. Nagamiya T, Nagata K, Kitano Y (1962) Magnetization process of a screw spin system. Prog Theor Phys 27(6):1253–1271

17. Kitano Y, Nagamiya T (1964) Magnetization process of a screw spin system II. Prog Theor Phys 31(1):1–43

18. Jensen J, Mackintosh AR (1990) Helifan: a new type of magnetic structure. Phys Rev Lett 64:2699–2702

19. Evenson WE, Liu SH (1968) Generalized susceptibilities and magnetic ordering of heavy rare earths. Phys Rev Lett 21:432–434

20. Evenson WE, Liu SH (1969) Theory of magnetic ordering in the heavy rare earths. Phys Rev 178:783–794

21. Andrianov AV (1995) Helical magnetic structures in heavy rare-earth metals as a probable manifestation of the electronic topological transition. J Magn Magn Mater 140-144:749–750. International Conference on Magnetism

22. Palmer SB, McIntyre GJ, Andrianov AV, Melville RJ (1998) The c/a ratio of helical magnetic structures in rare-earth materials. J Magn Magn Mater 177-181:1023–1024. International Conference on Magnetism (Part II)

23. Hughes ID, Däne M, Ernst A, Hergert W, Lüders M, Poulter J, Staunton JB, Svane A, Szotek Z, Temmerman WM (2007) Lanthanide contraction and magnetism in the heavy rare earth elements. Nature 446:650–653

24. Döbrich KM, Bostwick A, McChesney JL, Rossnagel K, Rotenberg E, Kaindl G (2010) Fermi-surface topology and helical antiferromagnetism in heavy lanthanide metals. Phys Rev Lett 104:246401

25. Andrianov AV, Savel'eva OA, Bauer E, Staunton JB (2011) Squeezing the crystalline lattice of the heavy rare-earth metals to change their magnetic order: experiment and ab initio theory. Phys Rev B 84:132401

26. Oroszlány L, Deák A, Simon E, Khmelevskyi S, Szunyogh L (2015) Magnetism of gadolinium: a first-principles perspective. Phys Rev Lett 115:096402

27. Liu SH (1978) Handbook of the physics and chemistry of rare Earths, vol 1. North-Holland, Amsterdam

28. Dugdale SB (2014) Probing the Fermi surface by positron annihilation and Compton scattering. Low Temp Phys 40(4):328–338

29. Zhou L, Wiebe J, Lounis S, Vedmedenko E, Meier F, Blügel S, Dederichs PH, Wiesendanger R (2010) Strength and directionality of surface Ruderman-Kittel-Kasuya-Yosida interaction mapped on the atomic scale. Nat Phys 6:187–191

30. Peters L, Ghosh S, Sanyal B, van Dijk C, Bowlan J, de Heer W, Delin A, Di Marco I, Eriksson O, Katsnelson MI, Johansson B, Kirilyuk A (2016) Magnetism and exchange interaction of small rare-earth clusters; Tb as a representative. Sci Rep 6:19676
31. Ohno H (1998) Making nonmagnetic semiconductors ferromagnetic. Science 281:951
32. Ali M, Adie P, Marrows CH, Greig D, Hickey BJ, Stamps RL (2007) Exchange bias using a spin glass. Nat Mater 6:70
33. Anisimov VI, Zaanen J, Andersen OK (1991) Band theory and Mott insulators: Hubbard U instead of Stoner I. Phys Rev B 44:943–954
34. Anisimov VI, Aryasetiawan F, Lichtenstein AI (1997) First-principles calculations of the electronic structure and spectra of strongly correlated systems: the LDA + U method. J Phys: Conden Matter 9(4):767
35. Lüders M, Ernst A, Däne M, Szotek Z, Svane A, Ködderitzsch D, Hergert W, Györffy BL, Temmerman WM (2005) Self-interaction correction in multiple scattering theory. Phys Rev B 71:205109
36. Perdew JP, Zunger A (1981) Self-interaction correction to density-functional approximations for many-electron systems. Phys Rev B 23:5048–5079
37. Hughes ID, Däne M, Ernst A, Hergert W, Lüders M, Staunton JB, Szotek Z, Temmerman WM (2008) Onset of magnetic order in strongly-correlated systems from ab initio electronic structure calculations: application to transition metal oxides. New J Phys 10(6):063010
38. Marmodoro A, Ernst A, Ostanin S, Sandratskii L, Trevisanutto PE, Lathiotakis NN, Staunton JB (2016) Short-range ordering effects on the electronic Bloch spectral function of real materials in the nonlocal coherent-potential approximation. Phys Rev B 94:224205
39. Langner MC, Roy S, Kemper AF, Chuang Y-D, Mishra SK, Versteeg RB, Zhu Y, Hertlein MP, Glover TE, Dumesnil K, Schoenlein RW (2015) Scattering bottleneck for spin dynamics in metallic helical antiferromagnetic dysprosium. Phys Rev B 92:184423
40. Teichmann M, Frietsch B, Döbrich K, Carley R, Weinelt M (2015) Transient band structures in the ultrafast demagnetization of ferromagnetic gadolinium and terbium. Phys Rev B 91:014425
41. Andres B, Christ M, Gahl C, Wietstruk M, Weinelt M, Kirschner J (2015) Separating exchange splitting from spin mixing in gadolinium by femtosecond laser excitation. Phys Rev Lett 115:207404
42. Krieger K, Dewhurst JK, Elliott P, Sharma S, Gross EKU (2015) Laser-induced demagnetization at ultrashort time scales: predictions of TDDFT. J Chem Theory Comput 11(10):4870–4874
43. Faulkner JS (1982) The modern theory of alloys. Prog Mater Sci 27(1):1–187
44. Schadler G, Weinberger P, Gonis A, Klima J (1985) Bloch spectral functions for complex lattices: applications to substoichiometric TiN_x and the Fermi surface of TiN_x. J Phys F: Metal Phys 15(8):1675
45. Callen HB, Callen E (1966) The present status of the temperature dependence of magnetocrystalline anisotropy, and the $l(l + 1)2$ power law. J Phys Chem Solids 27(8):1271–1285
46. McEwen KA (1978) Handbook on the physics and chemistry of rare Earths, vol 1. North-Holand, Amsterdam
47. Koehler WC, Cable JW, Child HR, Wilkinson MK, Wollan EO (1967) Magnetic structures of holmium II. the magnetization process. Phys Rev 158:450–461
48. Gebhardt JR, Ali N (1998) Magnetic phase diagrams of holmium determined from magnetoresistance measurements. J Appl Phys 83(11):6299–6301
49. Trushkevych O, Fan Y, Perry R, Edwards RS (2013) Magnetic phase transitions in $Gd_{64}Sc_{36}$ studied using non-contact ultrasonics. J Phys D: Appl Phys 46(10):105005
50. Locht ILM, Kvashnin YO, Rodrigues DCM, Pereiro M, Bergman A, Bergqvist L, Lichtenstein AI, Katsnelson MI, Delin A, Klautau AB, Johansson B, Di Marco I, Eriksson O (2016) Standard model of the rare earths analyzed from the Hubbard I approximation. Phys Rev B 94:085137
51. Rosen M (1968) Elastic moduli and ultrasonic attenuation of gadolinium, terbium, dysprosium, holmium, and erbium from 4.2 to 300K. Phys Rev 174:504–514
52. Palmer SB (1970) Debye temperatures of holmium and dysprosium from single crystal elastic constant measurements. J Phys Chem Solids 31(1):143–147

53. Bodryakov VY, Povzner AA, Zelyukova OG (1999) Magnetic contribution to the Debye temperature and the lattice heat capacity of ferromagnetic rare-earth metals (using gadolinium as an example). Phys Solid State 41(7):1138–1143
54. Gschneidner KA Jr, Pecharsky VK, Tsokol AO (2005) Recent developments in magnetocaloric materials. Rep Prog Phys 68(6):1479
55. Staunton JB, Marmodoro A, Ernst A (2014) Using density functional theory to describe slowly varying fluctuations at finite temperatures: local magnetic moments in gd and the 'not so local' moments of Ni. J Phys: Conden Matter 26(27):274210

Chapter 6
Frustrated Magnetism in Mn-Based Antiperovskite Mn₃GaN

The rich magnetism that originates in the family Mn_3AX (X = C and N, and A = Ni, Ag, Zn, Ga, In, Al, Sn, etc.) raises fundamental interest in the study of magnetic phase transitions and offers different pathways for their exploitation as magnetic refrigerants [1]. These metals crystallise into the cubic Mn-based antiperovskite structure, shown in Fig. 6.1a. The magnetism is due to Mn atoms positioned at the centre of the cube faces and the exchange interaction among them is mediated by the presence of the A and X atoms and their influence on the electronic glue. Depending on what elements are used very diverse magnetic phases can be obtained [1]. For example, the carbide's PM state (X = C, A = Zn, Al, Ga, In, Ge) is usually unstable to the formation of a second-order transition to a FM phase [1, 2] (Fig. 6.1b). The magnetism, however, becomes more complicated when the temperature is lowered, triggering transitions to other magnetic structures with antiferromagnetic ordering of different kind. The antiperovskite system that has possibly drawn more attention from this situation is Mn_3GaC [3–5], which has been subject of extensive experimental studies in the recent decades due to the significant inverse caloric response at the first-order transition from the FM state to an AFM order modulated by a wave vector $\mathbf{q} = (\frac{1}{2}, \frac{1}{2}, \frac{1}{2})\frac{2\pi}{a}$ [6–9] (Fig. 6.1c), where a is the lattice parameter. This is in sharp contrast with the physics observed in its counterpart Mn_3GaN, as well as in other nitrides in the family in general. In Mn_3GaN the moments in neighbouring Mn atoms interact antiferromagnetically leading to a geometrical magnetic frustration on the triangular lattice in the (111) planes. This results in a first-order transition from the PM state to a triangular AFM ordering, as found in neutron diffraction studies [2, 10]. Due to the lattice symmetries two different triangular phases are observed in experiment, corresponding to the Γ^{5g} (Fig. 6.1d) and Γ^{4g} (Fig. 6.1e) representations [10–12]. In both situations the three local magnetic moments on the Mn sites have equal sizes and their thermal fluctuations give rise to identical single-site magnetisations and angles of $2\pi/3$ between their directions, which yields a zero net magnetisation. The chirality in the Γ^{4g} and Γ^{5g} representations is the same and the difference lies in that the local moment directions differ by a simultaneous rotation of $\pi/2$ within the (111) plane, highlighted in colour blue in the figures. Hence, the energy difference between these two structures is purely attributed to spin-orbit coupling

© Springer Nature Switzerland AG 2020

E. Mendive Tapia, *Ab initio Theory of Magnetic Ordering*, Springer Theses,

https://doi.org/10.1007/978-3-030-37238-5_6

sources. Interestingly, this transition gains even more interest due to the negative thermal expansion associated with it [13, 14]. In fact, it is this anomaly the factor that renewed interest in these metallic compounds in the last decade. It is also worth mentioning that the stabilisation and competition between all the observed magnetic phases is often discussed in terms of signs and ratios among the nearest neighbour and the next nearest neighbour Mn-X-Mn and Mn-Mn magnetic interactions [4]. We show them as pairwise constants A_1, A_2 and A_3, respectively, in Fig. 6.1a. In this chapter we demonstrate, however, that multi-spin (or multi-site) interactions and their effect on the overall magnetic interactions play a fundamental role too.

If the cubic symmetry is distorted by strain one can expect the compromise among the magnetic interactions to be disrupted to trigger transitions to other magnetic states with net magnetisation [15]. Note that this is a different mechanism to magnetostriction, widely used in spintronic devices, which is driven by the more subtle relativistic spin-orbit coupling. Magnetic transitions driven by lattice strains have been reported in several perovskite oxides [16, 17]. One of the most important findings is a very strong dependence of the transition temperature on biaxial strain ($\approx$50 K per 1%) predicted [18] and subsequently confirmed experimentally [19] for G-type AFM phase of SrMnO$_3$. The ability to drive a magnetic phase transition with a large entropy change, $\Delta S = 9$ J/kgK, by means of biaxial strain was demonstrated in La$_{0.7}$Ca$_{0.3}$MnO$_3$ film on BaTiO$_3$ substrate [20]. Along these lines, magnetic frustration has been proposed to be an important phenomenon able to substantially enhance the caloric effect, as recently shown in the large barocaloric response measured at the transition in Mn$_3$GaN [21]. Our motivation is to study how the frustrated magnetism changes under biaxial strain and analyse the strain-temperature phase diagram in search of magnetic phase transitions that can be potentially used for cooling applications.

In this chapter we implement our SDFT-DLM theory, presented in Chaps. 2–4, to study the geometrically frustrated non-collinear magnetism of Mn$_3$GaN. We start in Sect. 6.1 by testing our results with available experimental data for the unstrained cubic system and give an estimation of the entropy change at the transition, suitably comparing with the barocaloric effect. Then we explore different tetragonal lattice distortions and study the elastocaloric effect in Sect. 6.2. We point out that our DLM method is well suited for this metallic system because the local moments are relatively well localised [22] while their interaction with the localised Ga p-states determines the size [15] and direction of the strain-induced total magnetisation, shown in $T = 0$ K theoretical studies.

6.1   Unstrained Cubic System

Due to the very complicated and diverse magnetism found experimentally in the Mn-based antiperovskite systems, we firstly carried out an analysis of the direct correlation function $\tilde{S}_{ss'}^{(2)}(\mathbf{q})$ of cubic Mn$_3$GaN. As explained in Sect. 3.4, the aim is to study the limit of the PM state ($\{\mathbf{m}_n\} \to 0$) and search for the most stable types of magnetic order. We, therefore, constructed fully disordered local moment potentials,

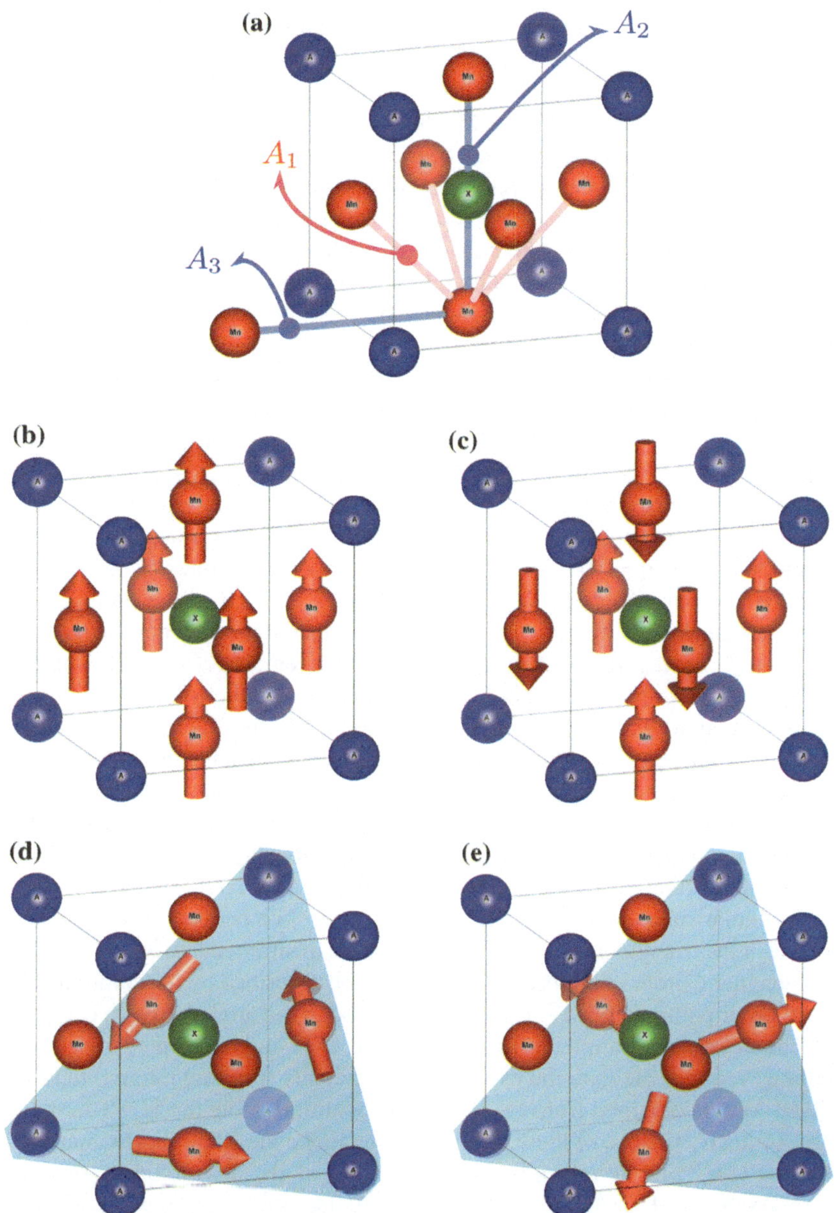

Fig. 6.1 **a** The antiperovskite paramagnetic (and also non-magnetic) unit cell of Mn_3AX systems and the magnetic interactions between nearest neighbours (red) and next nearest neighbours (blue). The element X sits on the cube centres and is surrounded by Mn atoms at the centre of the faces. The element A occupies the corner sites and can be one among many different elements or a solid solution of several (A = Ni, Ag, Zn, Ga, In, Al, Sn, etc.). The figure also shows some of the stable magnetic phases observed in experiment: The **b** ferromagnetic, **c** collinear antiferromagnetic (in Mn_3GaC), and the triangular antiferromagnetic in the **d** Γ_{5g} and **e** Γ_{4g} representations [2]. Arrows represent the averaged directions of the local magnetic moments and with colour blue we indicate the (111) plane

producing local moment sizes of $\mu \approx 3.1\mu_B$, and examined $\tilde{S}_{ss'}(\mathbf{q})$ in the reciprocal space set by them. Since there are three magnetic sites associated with the Mn atoms inside the PM unit cell (shown in Fig. 6.1a), diagonalising $\tilde{S}_{ss'}(\mathbf{q})$ yields three eigenfunctions $\tilde{u}_i(\mathbf{q})$ ($i = 1, 2, 3$). In Fig. 6.2a we show the largest value among them as a function of $\mathbf{q}$, i.e. the dependence of the most stable structure in the PM limit. We point out that the cubic lattice parameter a used is the one that minimises the total SDFT-based energy in our calculations. As we show in Fig. 6.2b, the value obtained is $a_0 = 4.14$ Å, which is approximately 6% larger than the experimentally measured of $a_{exp} = 3.898$ Å. We have found this significant lattice expansion indispensable for the appropriate description of the magnetism in Mn$_3$GaN. For example, the triangular state, corresponding to $\mathbf{q} = \mathbf{0}$ because there is no rotation of local order parameters from one unit cell to another (see Fig. 6.1d, e and Sect. 3.4), only stabilises when a is increased up to the minimum of the total energy. Indeed, the eigenvectors of $\max(\tilde{u}_i(\mathbf{0}))$ describe cosines of $2\pi/3$ angles between the local moments on the Mn sites at $a_0 = 4.14$ Å[1] but not at a_{exp}. However, inspecting Fig. 6.2a further reveals that, even though the triangular structure has a significantly large eigenvalue, in fact in the PM limit the system is unstable to the formation of another more stable magnetic phase modulated as $\mathbf{q} = (0.5\,0\,0.5)\frac{2\pi}{a}$, whose local order parameters, therefore, flip when moving along two crystallographic directions and remain the same along the third. Note that due to the lattice symmetries the calculation of $\max(\tilde{u}_i(\mathbf{q}))$ along (110) and (011) gives identical results to (101). The eigenvectors of this magnetic phase have only one non-zero component associated with the Mn atom from which when moving along (101) a N atom is not crossed. In other words, the calculation shows that the constant A_3 is large and negative. We advance that albeit the pairwise interactions studied now (enclosed inside $\tilde{S}_{ss'}(\mathbf{q})$) do not render the triangular phase at the free energy minimum, the presence of multi-site interactions in fact stabilise the triangular ordering inducing a first-order transition. In the following we show this from the calculation of the internal magnetic fields at large magnetic orderings of the triangular state. We refer again back to Chap. 4 for a deeper discussion on this.

In case of the triangular state in Mn$_3$GaN with an unstrained lattice (cubic symmetry) the magnetic order of all three Mn atoms can be described by the common size of the local order parameters, m_{Tr} (or λ_{Tr}), even though the angle between each pair is $2\pi/3$. The situation is completely analogous to the case study presented in Sect. 4.2.2. Here, however, we calculate the expansion coefficients from the internal field $\{\mathbf{h}_n^{int}\}$ SDFT-DLM data obtained from the PM potentials minimising the SDFT-based total energy. To justify the use of these potentials we invoke the RSA approximation. We set $\{\boldsymbol{\lambda}_n\} = \{\beta \mathbf{h}_n^{int}\}$ input values for the calculation of $\{\mathbf{h}_n^{int}\}$ with values ranging as $\lambda_{Tr} = 0.05, 0.1, 0, 1, 2, 3, 4, 5, 6, 8$, and 10, satisfactorily sampling the m_{Tr} space. We have found that $\{\mathbf{h}_n^{int}\}$ is described very well by the two term equation

$$\langle \Omega^{int} \rangle_0 = -a_{Tr} m_{Tr}^2 - b_{Tr} m_{Tr}^4, \tag{6.1}$$

---

[1] An example of the eigenvector components of $\max(\tilde{u}_i(\mathbf{0}))$ describing the triangular phase in one of our calculations is for example $(0.674, -0.736, 0.616)$.

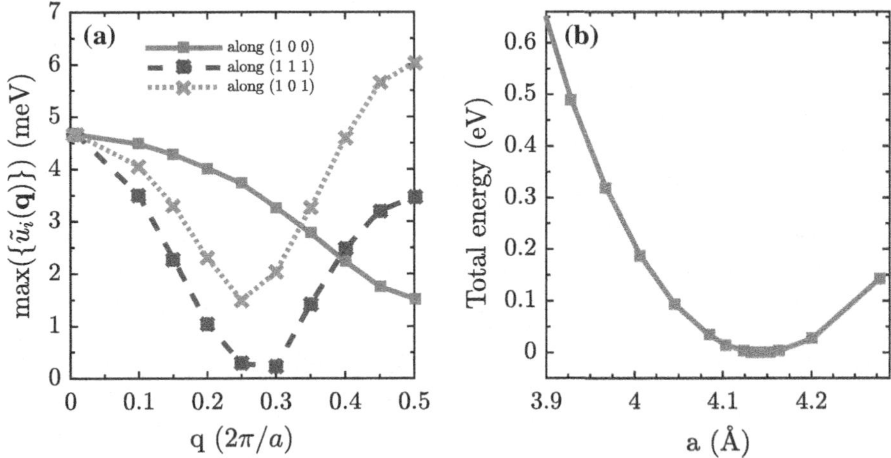

Fig. 6.2 Panel **a** shows the dependence of the maximum eigenvalue of the direct correlation function $\tilde{S}_{ss'}(\mathbf{q})$ on the reciprocal wave vector $\mathbf{q}$ for a lattice parameter minimising the SDFT-based total energy $a_0 = 4.14\,\text{Å}$, which is plotted in panel **b**. The results are shown for three characteristic directions from the centre to the boundaries of the Brillouin zone, red squares for (100), blues stars for (111), and green crosses for (101). The triangular state experimentally observed in Mn_3GaN corresponds to $\mathbf{q} = \mathbf{0}$

where the coefficients are $a_{Tr} = 31.47$ meV and $b_{Tr} = 33.11$ meV. Note that these constants contain the effect of the pairwise interactions, including A_1, A_2, and A_3, and the four-site constants, respectively, as illustrated in Eq. (4.12). We proceed by constructing the free energy per formula unit as prescribed by Eq. (3.26), which under the presence of no external magnetic field and magnetic entropies associated with the three Mn atoms, S_1, S_2 and S_3, defined by Eq. (3.24), gives

$$\mathcal{G}_1 = -a_{Tr}m_{Tr}^2 - b_{Tr}m_{Tr}^4 - T(S_1 + S_2 + S_3). \tag{6.2}$$

After minimising Eq. (6.2) at different values of the temperature our SDFT-DLM approach predicts the triangular AFM state as the most stable structure at high temperatures. At the Néel temperature the quartic coefficient b_{Tr} is larger than its positive counterpart in the expansion of magnetic entropy in powers of the order parameter (see Eq. (4.8)), which indicates a first-order phase transition between the triangular AFM and the PM state in agreement with experiment [21], and therefore stabilising this magnetic state at higher temperature than the $\mathbf{q} = (0.5\ 0\ 0.5)\frac{2\pi}{a}$ phase. We find $T_N = 304$ K, which is very close to reported experimental values $T_N = 288$ K [13] and 290 K [21]. When we repeat our simulation for a slightly larger lattice parameter, preserving the cubic symmetry and the form of Eq. (6.2), we find an increase of T_N with increasing the unit cell volume, in agreement with experiment and hence capturing the negative thermal expansion [14, 21]. Matsunami et al. have recently measured a very large isothermal entropy change

$\Delta S = 22.3$ J/kgK at the AFM-PM transition in Mn$_3$GaN [21] which correlates with the large magnetovolume effect [13] driven by the geometric frustration and the abrupt change of effective amplitudes of Mn magnetic moments. Our DLM theory for unstrained cubic Mn$_3$GaN finds a significantly larger barocaloric response (Fig. 2 of Ref. [21]), estimated from the change of $S_{\text{mag}} = S_1 + S_2 + S_3$ yielding at T_N the value $\Delta S_{\text{mag}} = S_{\text{mag}}(T_N + \delta T) - S_{\text{mag}}(T_N - \delta T) = 103.2$ J/kgK, accompanied by a large change of the magnetic order parameter $\Delta m_{Tr} = 0.74$. The electronic entropy change calculated from the dependence of the density of states on m_{Tr} and Eq. (4.25) is very low, $\Delta S_{el} = 0.035$ J/kgK, and the vibrational contribution estimated from the Debye model introduced in Eq. (4.25), ΔS_{vib} (we have assumed a Debye temperature of 429.2 K, which becomes after rescaling to our transition temperature [13]), cannot compensate the discrepancy between ΔS_{mag} and the measured ΔS_{tot} either [21]. However, it should be noted that our ΔS_{mag} falls well below the theoretical upper limit proportional to $k_B \ln(2J + 1)$ [23], which is 161.52 J/kgK for Mn$_3$GaN, where J is the total angular momentum of the magnetic atom. At the same time, strong dependence of magnetic transitions on compositional disorder has been shown in Mn$_3$AN [13, 14] and FeRh [24]. Owing to its ties to the geometric frustration, ΔS_{mag} is likely to be sensitive also to any symmetry lowering due to structural defects in polycrystalline samples. Therefore, ΔS_{mag} calculated in a system with perfect stoichiometry and lattice symmetry, with a very sharp phase transition in consequence, should be regarded as an upper estimate of the entropy change measured at a smoother phase transition in a real sample. This is consistent with our overestimate of ΔS_{mag} and indicates that the simulated material has a sharper phase transition than the available sample.

6.2   Biaxial Strain, the Elastocaloric Effect and Cooling Cycles

Having compared the results of our SDFT-DLM modelling with available experimental data for Mn$_3$GaN with cubic symmetry, we now focus on the effect of tetragonal lattice distortions. We assume only volume-conserving strains, i.e. Poisson's ratios equal to 0.5. Highly pertinent to the physics investigated in this section is the work carried out in Ref. [15]. The authors studied the effect of net magnetisation induced by elastic strain in frustrated AFM ordering in a long range of A elements in Mn$_3$AN from zero temperature SDFT calculations. Their calculations show that an applied biaxial strain $\varepsilon_{xx} = \varepsilon_{yy} = (a - a_0)/a_0 \neq \varepsilon_{zz}$ (where a_0 is the lattice parameter of the relaxed structure) has a large impact on the frustration and leads to canting and relative change of uncompensated local moment sizes in a distorted triangular state, which we show in panels (a) and (b) of Fig. 6.3 for compressive and tensile strains in Mn$_3$GaN, respectively. Our results in this section are in agreement with these insights and expand the study to non-zero temperature.

(a) Compressive - Canted triangular   (b) Tensile - Canted triangular

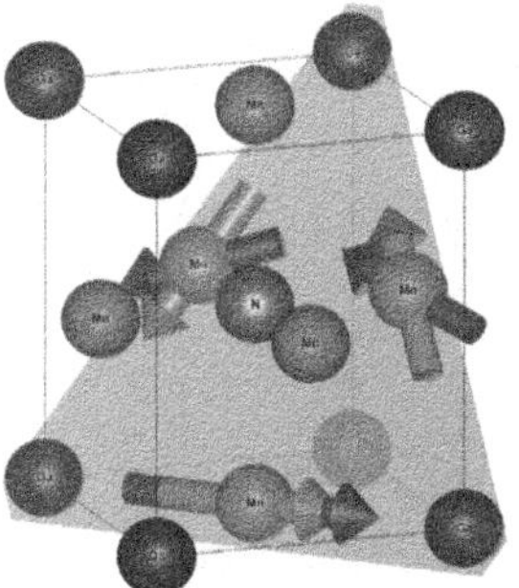 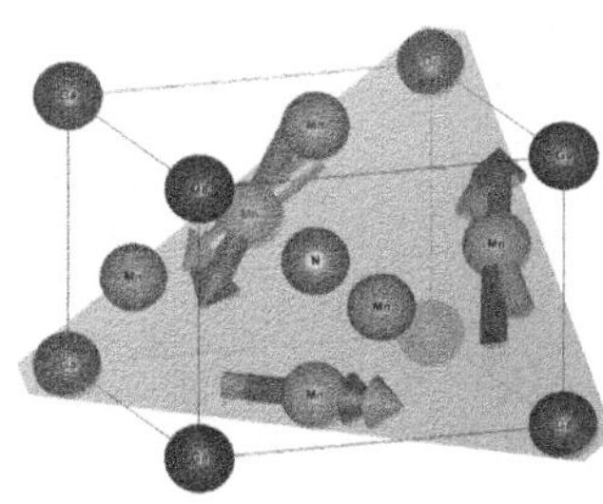

(c) Compressive - Collinear ferrimagnetic   (d) Tensile - Collinear antiferromagnetic

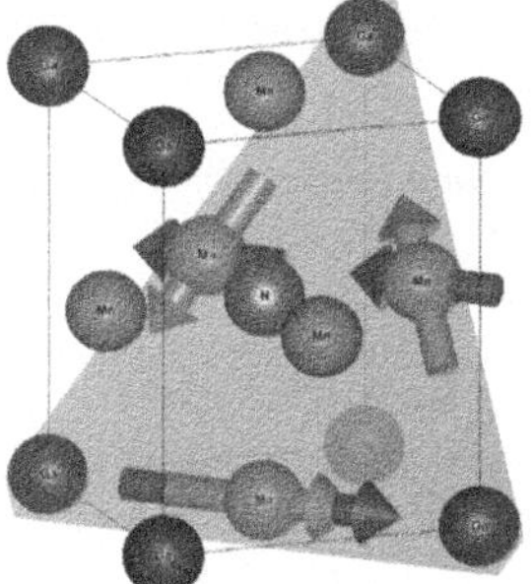 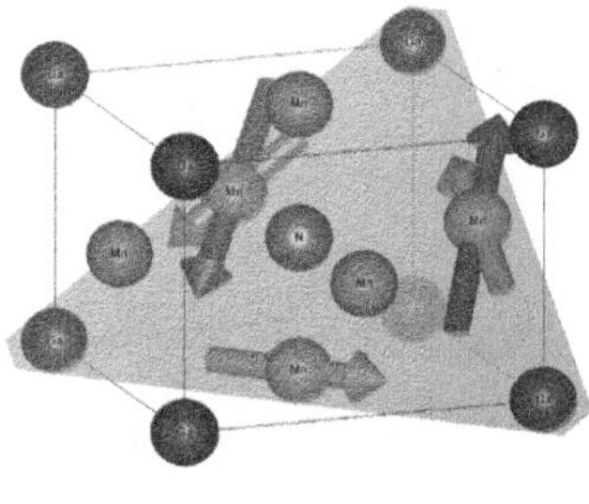

Fig. 6.3 The structure of Mn_3GaN under compressive and tensile biaxial strains. Magenta arrows are used to represent the induced canted angles and changes of the local order parameters and magnetic moment order sizes. These changes as well as the tetragonal distortions are exaggerated for illustrative purposes. Red arrows show the unstrained magnetic structure for comparison (see Fig. 6.1d to distinguish between strained and unstrained arrows). Upper and lower panels show the results at low and high temperatures, respectively: The **a**, **b** canted triangular state, and collinear **c** ferrimagnetic and **d** antiferromagnetic orderings. For compressive/tensile strain the magnetic order parameter (represented by the size of the arrow) of the Mn atom at lattice position (0.5 0.5 0) increases/decreases. For the collinear antiferromagnetic state the magnetic order at this sub-lattice becomes fully frustrated and so it is supressed

We perform our finite temperature study by firstly examining how the eigenfunctions $\{\tilde{u}_i(\mathbf{q})\}$, and the corresponding eigenvectors, of the direct correlation function change against different values of ε_{xx} in the PM state. In Fig. 6.4 we show the largest eigenfunction, $\tilde{u}_p(\mathbf{q}) = \max(\{\tilde{u}_i(\mathbf{q})\})$, among the three possible solutions $i = 1, 2, 3$. Clearly, both positive and negative stresses strongly favour the triangular state, now distorted, at $\mathbf{q} = \mathbf{0}$, in clear detriment of the $\mathbf{q} = (0.5\ 0\ 0.5)\frac{2\pi}{a}$ modulated phase and ensuring the stability, therefore, of a magnetic ordering prescribed by $\mathbf{q} = \mathbf{0}$. We emphasise that the change of the eigenfunction at this value is remarkably large. In terms of the associated transition temperature, calculated using Eq. (3.76), our results predict an increase of approximately 70 K for a relatively small 1% tetragonal distortion. This is in line with the behaviour of the eigenvectors. An inspection of their

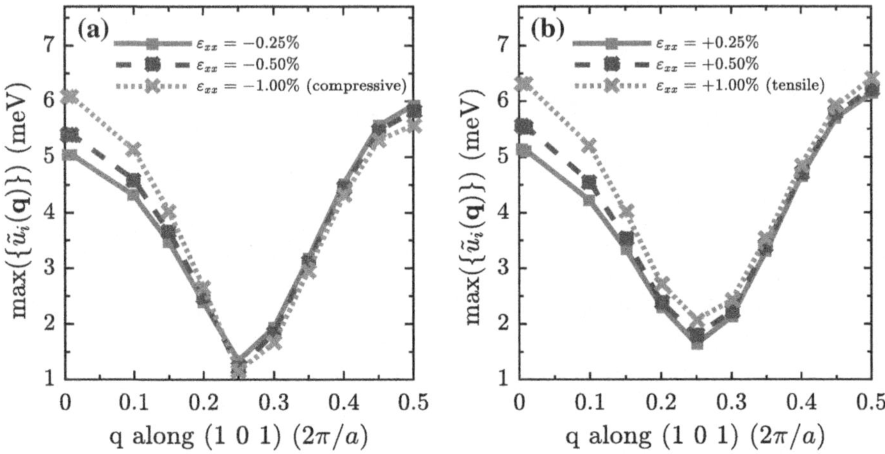

Fig. 6.4 The dependence of the largest value among $\{\tilde{u}_i(\mathbf{q}),\, i = 1, 2, 3\}$ on the reciprocal wave vector $\mathbf{q}$ along the relevant direction (101). The results are shown for different values of **a** compressive ($\varepsilon_{xx} < 0$) and **b** tensile ($\varepsilon_{xx} > 0$) biaxial strains

Table 6.1 The eigenvectors ($V_{p,1}(\mathbf{q} = \mathbf{0})$, $V_{p,2}(\mathbf{q} = \mathbf{0})$, $V_{p,3}(\mathbf{q} = \mathbf{0})$) associated to $\tilde{u}_p(\mathbf{q} = \mathbf{0}) = \max(\tilde{u}_i(\mathbf{q} = \mathbf{0}))$ for the same compressive and tensile strains shown in Fig. 6.4. The components correspond to the three Mn atoms in the PM unit cell illustrated in Fig. 6.1(a). In particular, $V_{p,1}$ and $V_{p,2}$ are linked to the atomic positions (0 0.5 0.5) and (0.5 0 0.5), and $V_{p,3}$ to (0.5 0.5 0), in lattice parameter units

	Compressive ($\varepsilon_{xx} < 0$)	Tensile ($\varepsilon_{xx} > 0$)
$|\varepsilon_{xx}| = 0.25\%$	(0.392, 0.392, −0.832)	(1, 1, 0)
$|\varepsilon_{xx}| = 0.50\%$	(0.378, 0.378, −0.845)	(1, 1, 0)
$|\varepsilon_{xx}| = 1.00\%$	(0.350, 0.350, −0.869)	(1, 1, 0)

components $\{(V_{p,1}(\mathbf{q} = \mathbf{0}),\, V_{p,2}(\mathbf{q} = \mathbf{0}),\, V_{p,3}(\mathbf{q} = \mathbf{0}))\}$ (see Sect. 3.4 and specifically Eqs. (3.73) and (3.80) for their meaning), shown in Table 6.1, reveals that even the smallest studied distortion of 0.25% fundamentally affects their shape, which changes from one describing a compensated triangular state to a different one that remains the same for larger stresses. In particular, compressive and tensile strains automatically induce ferrimagnetic (FIM) and collinear AFM orderings described by eigenvectors $(+, +, −)$ and $(+, −, 0)$, respectively. Hence, the instability of the PM state at high temperatures is rapidly increased under strain application to the formation of these two previously unreported magnetic structures. For clarity, we remark that a component with value zero, as in the collinear AFM eigenvector, indicates a fully disordered magnetic site, caused by thermal fluctuations of the orientation of a local moment with size μ. It is regarded, therefore, as a entirely magnetically frustrated site in which all magnetic directions are equally probable. The direct consequence of this is that whilst tensile strains increase the magnetic frustration, compressive strains release it.

We now proceed by generalising $\langle \Omega^{\text{int}} \rangle_0$, given in Eq. (6.1), to account for the distorted triangular state and the emerging FIM and collinear AFM magnetic phases found by the analysis of the direct correlation function. A single order parameter is not sufficient to describe the canted magnetism of the lower symmetry caused by the tetragonal distortion. Thus, we define three independent order parameters $\{\mathbf{m}_1, \mathbf{m}_2, \mathbf{m}_3\}$ for each magnetic atom inside the unit cell. Note that, as indicated in the caption of Table 6.1, index 3 is used for the Mn atom in the (0.5 0.5 0) position without a strain-induced local moment rotation, and the indices 1 and 2 for the other two magnetic sites lying in the same (0 0 1) plane. We produced SDFT-DLM data of $\{\mathbf{h}_n^{\text{int}}\}$ from input values of $\{\mathbf{m}_1, \mathbf{m}_2, \mathbf{m}_3\}$ thoroughly sampling the FIM, collinear AFM, and distorted triangular phases. The calculation was repeated for PM self-consistent potentials obtained after biaxial strain applications ranging as $\varepsilon_{xx} = 0, \pm 0.25\%, \pm 0.50\%, \pm 1.00\%$. Evidently, the SDFT-DLM internal energy fitting this data is more complicated than Eq. (6.1) capturing the effect of the tetragonal symmetry,

$$
\begin{aligned}
\langle \Omega^{\text{int}} \rangle_0 = &- a_1 (m_1^2 + m_2^2) - a_2 m_3^2 - \alpha_1 \mathbf{m}_3 \cdot (\mathbf{m}_1 + \mathbf{m}_2) - \alpha_2 \mathbf{m}_1 \cdot \mathbf{m}_2 \\
&- b_1 (m_1^4 + m_2^4) - b_2 m_3^4 - \beta_1 \left[(\mathbf{m}_3 \cdot \mathbf{m}_1) m_2^2 + (\mathbf{m}_3 \cdot \mathbf{m}_2) m_1^2 \right] \\
&- \beta_2 \left[(\mathbf{m}_3 \cdot \mathbf{m}_1) + (\mathbf{m}_3 \cdot \mathbf{m}_2) \right] (\mathbf{m}_1 \cdot \mathbf{m}_2) - \beta_3 (\mathbf{m}_1 \cdot \mathbf{m}_2) m_3^2 \\
&- \beta_4 (\mathbf{m}_3 \cdot \mathbf{m}_1)(\mathbf{m}_3 \cdot \mathbf{m}_2).
\end{aligned}
\tag{6.3}
$$

Including only quadratic and quartic terms in Eq. (6.3) is enough to fit satisfactorily $\{\mathbf{h}_n^{\text{int}}\}$ across the explored range of strain and local order parameter sampling. As a consequence, the free energy $\mathcal{G}_1 = \langle \Omega^{\text{int}} \rangle_0 - T(S_1 + S_2 + S_3)$ now contains the complexity of Eq. (6.3) and its minimisation renders the most stable magnetic structure among the magnetic phases analysed, i.e. distorted triangular, FIM, and collinear AFM. In order to construct temperature-strain phase diagrams we have also studied the behaviour of the quadratic, $\{a_1, a_2, \alpha_1, \alpha_2\}$, and quartic, $\{b_1, b_2, \beta_1, \beta_2, \beta_3, \beta_4\}$, constants for the seven different values of $\varepsilon_{xx} \in \langle -1, 1 \rangle \%$ explored. The quadratic terms follow a very good linear dependence, as shown in Fig. 6.5. Moreover, the change of the quartic coefficients with ε_{xx} has been found to have a very small effect on the phase diagram. Keeping them constant at their values for the relaxed cubic system, $b_1 = b_2 = 2.856\,\text{meV}$, $\beta_1 = \beta_3 = -37.54\,\text{meV}$, and $\beta_2 = \beta_4 = 43.08\,\text{meV}$, produces no qualitative and nearly negligible quantitative changes in the transition temperatures. We, therefore, kept them constant and only performed an additional linear fitting of the quadratic constants to extract their dependence on ε_{xx}. In addition, we obtained a linear behaviour of the Mn magnetic moment sizes described as $\mu_1 = \mu_2 = (3.102 - 0.0361\varepsilon_{xx})\mu_B$ and $\mu_3 = (3.102 + 0.0438\varepsilon_{xx})\mu_B$ for compressive strain, and $\mu_1 = \mu_2 = (3.102 - 0.0265\varepsilon_{xx})\mu_B$ and $\mu_3 = (3.102 + 0.0444\varepsilon_{xx})\mu_B$ for tensile strain.

In Fig. 6.6 we plot the phase diagram constructed by tracking the free energy of competing magnetic phases across a range of strain and temperature with a sufficiently small step allowed by the fitting for the dependence on ε_{xx}. Our finite tem-

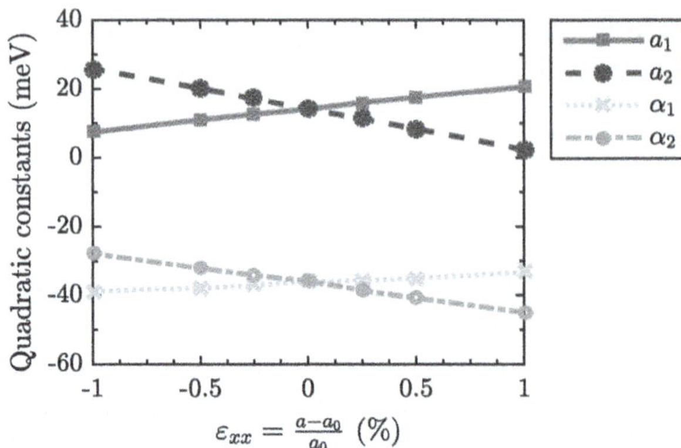

Fig. 6.5 The linear dependence of the quadratic constants a_1, a_2, α_1, and α_2 obtained from fitting the SDFT-DLM data of $\{\mathbf{h}_n^{\text{int}}\}$ for $\varepsilon_{xx} = 0$, $\pm 0.25\%$, $\pm 0.50\%$, $\pm 1.00\%$

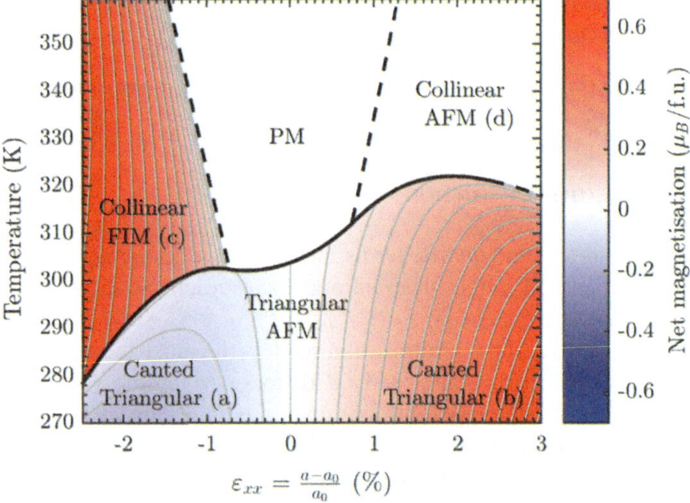

Fig. 6.6 Temperature-strain magnetic phase diagram of Mn₃GaN. Colours encodes the size and orientation of the induced net magnetic moment, $\boldsymbol{\mu}_{net}$, along the (110) axis. Thick black lines indicate first-order (solid) and second-order (dashed) magnetic phase transitions. Letters in brackets link to panels of Fig. 6.3

perature simulations are in perfect agreement with the results reported in Ref. [15] at $T = 0$ K. At low temperatures we predict the same sign of the canting in Mn₃GaN, as indicated in Fig. 6.3a, b. A net magnetisation is induced along the (110) axis, which changes sign when switching between compressive and tensile strains. In agreement with the analysis of the PM state, at higher temperatures we predict two novel strain-induced magnetic phases, a collinear FIM phase for compressive strain $\varepsilon_{xx} < -0.75\%$ and a collinear AFM phase for tensile strain $\varepsilon_{xx} > 0.75\%$, shown in

Fig. 6.3c and 6.3d, respectively. Remarkably, we also determine the order of the phase transitions. Solid (dashed) black lines indicate first- (second-) order transitions. We point out that the collinear FIM and AFM phases emerge from the large change of the quadratic coefficients with ε_{xx}. This is the most conspicuous feature of Fig. 6.6, leading to a strong dependence of the second-order transition temperatures, between these phases and the PM state, on ε_{xx}. The transition between canted triangular and collinear AFM states changes from first- to second-order for large tensile strain, generating a tricritical point in consequence. The colour-coding shows the net magnetisation $\boldsymbol{\mu}_{net} = \mathbf{m}_1\mu_1 + \mathbf{m}_2\mu_2 + \mathbf{m}_3\mu_3$ per unit cell. For example, the collinear AFM state does not possess any net magnetic moment, whereas the tensile-strained canted triangular and collinear FIM states develop $\boldsymbol{\mu}_{net}$ parallel to (110) (positive, red). The compressive-strained canted triangular state, however, has $\boldsymbol{\mu}_{net}$ antiparallel to (110) (negative, blue). From this follows another important prediction related to the compressive region of the phase diagram ($\varepsilon_{xx} < 0$). The transition from canted triangular to the FIM state involves a full reversal of the net magnetisation, which can be triggered both thermally and/or by changing ε_{xx}. If straining by means of a device based on a Mn_3GaN film deposited epitaxially on a piezoelectric substrate, this region of the phase diagram could be used for repeatable magnetisation reversal under purely electrical control of ε_{xx}. Such a device would not require assisting with an external magnetic, which is the outstanding goal in magnetoelectrics. Regarding other relevant traits of the phase diagram, the corresponding induced magnetic field reaches 200 Oe at 1% strain at room temperature. The strained material, therefore, has a potential for multicaloric effects tuned and/or driven by a magnetic field. To finalise the discussion of Fig. 6.6, we remark that the only relevant dependence on ε_{xx} is the one attributed to the quadratic coefficients. Hence, we concluded that all features of the temperature-strain magnetic phase diagram are determined mainly by two factors. (i) The presence of large quartic coefficients giving rise to the first-order nature of the PM-AFM transition at zero strain and (ii) a strong dependence of the quadratic coefficients on ε_{xx}.

Figure 6.7 presents our magnetic phase diagram from the perspective of the total entropy. The electronic contribution S_{el} has been ignored because we have found it to be negligible compared with S_{mag} and S_{vib}, as explained in Sect. 6.1. This figure shows the abrupt entropy change $\Delta S_{tot} \approx 100$ J/kgK at the first-order transition to the PM state at zero strain. The transition gradually becomes less first-order-like as the strain increases and becomes smooth around $\varepsilon_{xx} = 2.5\%$ due to the presence of the tricritical point. Both the collinear FIM and AFM states show a very strong dependence of S_{mag} on strain at higher temperatures.

We now propose an elastocaloric cooling cycle utilising the complex pattern of magnetic phase transitions of Fig. 6.7, which can be considered as an alternative to structural phase transitions, in shape memory alloys for example. The cycle starts by adiabatic application of strain (red line from point (1) to (2)), which causes a decrease of S_{mag} at the second-order transition from the PM to collinear AFM state. This entropy change is compensated by an increase in S_{vib} accompanied by a warming of approximately 25 K. In the next step the system is cooled to its original temperature at constant strain (black line from point (2) to (3)). At this point S_{mag}

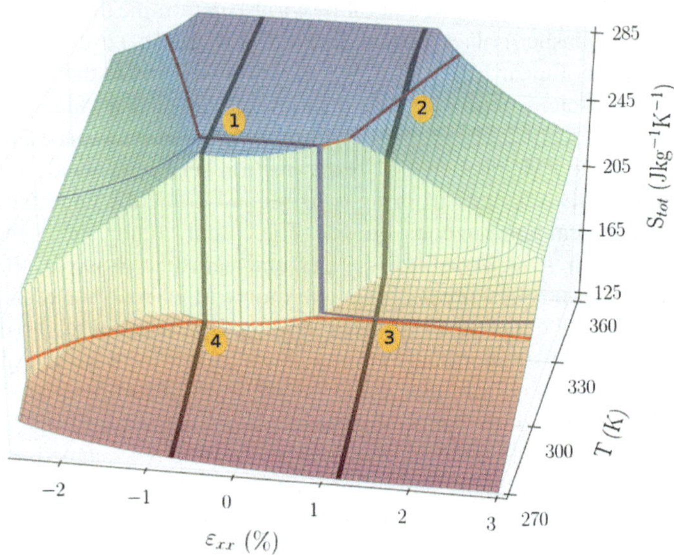

Fig. 6.7 Total entropy of Mn$_3$GaN against the temperature and biaxial strain. Red contour lines mark adiabatic application of strain at $S_{tot} = 170$ and 270 J/kgK. Black lines mark iso-strain cooling ($\varepsilon_{xx} = 1.18\%$) and heating ($\varepsilon_{xx} = -0.73\%$). Blue isotherm marks the reference temperature of 308 K and orange numbers indicate the proposed cooling cycle

further decreases through the first-order transition to the canted triangular state and heat is expelled to the environment. In the third step a strain is applied adiabatically again (red line from point (3) to (4)) entailing a continuous increase of S_{mag} and a temperature decrease of approximately 5 K. Finally, the refrigerant is warmed up at constant strain to close the cycle (black line from point (4) back to (1)). S_{mag} increases sharply at the first-order transition from the canted triangular state to the PM state and heat is absorbed from the load.

Figure 6.8 shows the dependence of S_{tot} on temperature for five relevant strains, which determine the maximum isothermal entropy change ΔS_{iso}^{max} and different values of the largest adiabatic temperature change ΔT_{ad}^{max}. We recall that in case of a cooling cycle with a single first-order phase transition driven by an external magnetic field using a material with a weak dependence of the Curie temperature on field its ΔT_{ad}^{max} cannot reach the highest value allowed by the entropy change [25, 26]. In our case, the rate of change of T_N with strain is relatively small compared to the large ΔS_{iso}^{max} in Mn$_3$GaN which would limit $\Delta T_{ad_1}^{max}$ if the elastocaloric-based cooling cycle was restricted to strains below 0.75%, as indicated in Fig. 6.8. However, at larger strains the cooling cycle benefits from the additional second-order transition between the collinear magnetic structures and the PM state with remarkably high dependence on ε_{xx}. This causes a previously unreported qualitative change of the temperature dependence of $S_{tot}(\varepsilon, T)$. For large enough values of ε_{xx} the collinear structures are stabilised and two phase transitions are triggered with increasing temperature,

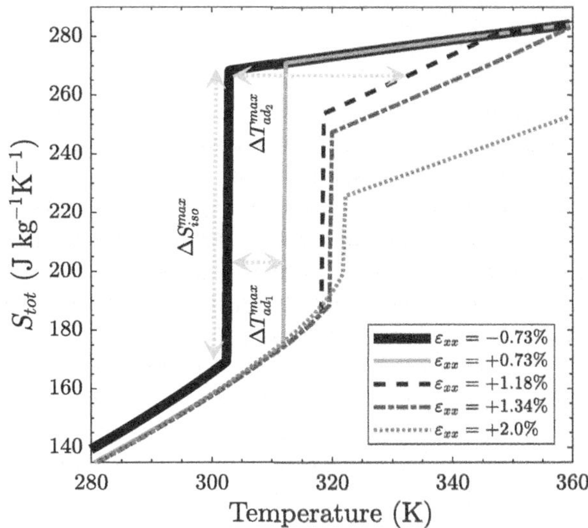

Fig. 6.8 The total entropy for selected values of strain in Mn$_3$GaN. Continuous lines correspond to iso-strain curves where only the first-order phase transition is crossed (small $\Delta T_{ad_1}^{max}$). All dashed, dash-dotted, and dotted lines cross both the first- and second-order phase transitions allowing to access large $\Delta T_{ad_2}^{max}$

namely first-order canted triangular-to-collinear FIM (or AFM) and second-order collinear FIM (or AFM)-to-PM. Dashed, dash-dotted, and dotted lines in Fig. 6.8 show this situation. As a result, the adiabatic temperature change is substantially increased from $\Delta T_{ad_1}^{max}$ to $\Delta T_{ad_2}^{max}$. Hence, our elastocaloric cycle offers simultaneously both large $\Delta S_{iso}^{max} \approx 100$ J/kgK and $\Delta T_{ad_2}^{max} \approx 30$ K. Even if the corresponding experimental ΔS_{iso}^{max} was a factor of 5 lower (as suggested by the observed barocaloric effect [21]) the proposed cycle would still be highly competitive with the available magnetocaloric and mechanocaloric counterparts [25, 27]. We conclude that the combination of both first- and second-order transitions improves significantly the cooling capacity of the elastocaloric cycle. We stress that the stability of the collinear magnetic structures is underpinned by the strong spin-lattice coupling arising from the frustrated exchange interactions and the effect of biaxial strain on them. The transition temperatures are in the room temperature range and, in principle, they can be further tuned by partially substituting atom A by an element with a different number of valence s- and p-electrons [2, 13]. In addition, the availability of phase transitions between two ordered states is relevant for elastocaloric-based cooling applications as it can reduce losses due to spin fluctuations and short-range order of a PM state [28].

6.3   Summary and Conclusions

In this chapter we have modelled the geometrically frustrated magnetism of non-collinear and collinear magnetic phases in Mn-antiperovskite nitride Mn$_3$GaN with relaxed and biaxially strained lattice at finite temperatures. We have applied the SDFT-DLM theory set out in Chaps. 2–4 to investigate the effect of changing temperature on the magnetic ordering and elastocaloric responses. The study has been structured by firstly analysing the direct correlation function in the paramagnetic limit and subsequently using the results obtained to establish the more extensive calculation of the internal magnetic fields at different magnetic orderings. Prior to the study of tetragonal distortions, we have started by investigating the relaxed cubic state under ambient conditions. We found the stability of the triangular phase at low temperature and a first-order transition to the PM phase at a Néel temperature in good agreement with available experimental data. The theory is able to provide the relevant thermodynamic quantities to describe the stability of competing magnetic phases, which allowed us to construct a strain-temperature magnetic phase diagram. We predict two novel magnetic phases, namely collinear ferrimagnetic at $\varepsilon_{xx} <$ 0.75% and collinear antiferromagnetic at $\varepsilon_{xx} > 0.75\%$. These magnetic structures arise from the increase and decrease of magnetic frustration caused by the strain, respectively. They are stable at high temperatures and show a second-order transition to the PM state which strongly depends on biaxial strain. The combination of both second-order and first-order transitions enabled us to propose an elastocaloric cooling cycle which exhibits large isothermal entropy change and adiabatic temperature change simultaneously. This rich phenomenology is available due to the strong spin-lattice coupling linked to the magnetic frustration. We finally remark that the Mn$_3$AN family of frustrated non-collinear AFMs might offer opportunities to tune the chemical composition to achieve better cooling performance than shown here, while still utilising relatively abundant chemical elements. We thus suggest Mn$_3$AN as a new class of elastocaloric materials.

References

1. L'Héritier Ph, Fruchart D, Madar R, Fruchart R (1988) Alloys and compounds of d-elements with main group elements. Part 2. In: Landolt-Börnstein - Group III condensed matter Wijin HPJ (ed) New Series III/19c. Springer, Berlin
2. Fruchart D, Bertaut EF (1978) Magnetic studies of the metallic perovskite-type compounds of manganese. J Phys Soc Jpn 44(3):781–791
3. Bouchaud J-P, Fruchart R, Pauthenet R, Guillot M, Bartholin H, Chaissé F (1966) Antiferromagnetic-ferromagnetic transition in the compound Mn$_3$GaC. J Appl Phys 37(3):971–972
4. Fruchart D, Bertaut EF, Sayetat F, Nasr Eddine M, Fruchart R, Sénateur JP (1970) Structure magnetique de Mn$_3$GaC. Solid State Commun 8(2):91–99
5. Shirai M, Ōhata Y, Suzuki N, Motizuki K (1993) Electronic structure and ferromagnetic-antiferromagnetic transition in cubic perovskite-type compound Mn$_3$GaC. Jpn J Appl Phys 32(S3):250
6. Yu M-H, Lewis LH, Moodenbaugh AR (2003) Large magnetic entropy change in the metallic antiperovskite Mn$_3$GaC. J Appl Phys 93(12):10128–10130
7. Tohei T, Wada H, Kanomata T (2003) Negative magnetocaloric effect at the antiferromagnetic to ferromagnetic transition of Mn$_3$GaC. J Appl Phys 94(3):1800–1802
8. Çakir Ö, Acet M (2012) Reversibility in the inverse magnetocaloric effect in Mn$_3$GaC studied by direct adiabatic temperature-change measurements. Appl Phys Lett 100(20):202404
9. Çakir Ö, Acet M, Farle M, Senyshyn A (2014) Neutron diffraction study of the magnetic-field-induced transition in Mn$_3$GaC. J Appl Phys 115(4):043913

10. Bertaut EF, Fruchart D, Bouchaud JP, Fruchart R (1968) Diffraction neutronique de Mn_3GaN. Solid State Commun 6(5):251–256

11. Fruchart D, Bertaut EF, Madar R, Lorthioir G, Fruchart R (1971) Structure magnetique et rotation de spin de Mn_3NiN. Solid State Commun 9(21):1793–1797

12. Meimei W, Wang C, Sun Y, Chu L, Yan J, Chen D, Huang Q, Lynn JW (2013) Magnetic structure and lattice contraction in Mn_3NiN. J Appl Phys 114(12):123902

13. Takenaka K, Ichigo M, Hamada T, Ozawa A, Shibayama T, Inagaki T, Asano K (2014) Magnetovolume effects in manganese nitrides with antiperovskite structure. Sci Technol Adv Mater 15(1):015009

14. Takenaka K, Takagi H (2005) Giant negative thermal expansion in Ge-doped anti-perovskite manganese nitrides. Appl Phys Lett 87(26):261902

15. Zemen J, Gercsi Z, Sandeman KG (2017) Piezomagnetism as a counterpart of the magnetovolume effect in magnetically frustrated Mn-based antiperovskite nitrides. Phys Rev B 96:024451

16. Callori SJ, Hu S, Bertinshaw J, Yue ZJ, Danilkin S, Wang XL, Nagarajan V, Klose F, Seidel J, Ulrich C (2015) Strain-induced magnetic phase transition in $SrCoO_{3-}$ thin films. Phys Rev B 91:140405

17. Ma C-L, Dai C-M, Chen G-Y, Chen D, Zang T-C, Ge L-J, Zhou W, Zhu Y (2015) Strain effect on the Néel temperature of $SrTcO_3$ from first-principles calculations. Solid State Commun 219:25–27

18. Lee JH, Rabe KM (2010) Epitaxial-strain-induced multiferroicity in $SrMnO_3$ from first principles. Phys Rev Lett 104:207204

19. Maurel L, Marcano N, Prokscha T, Langenberg E, Blasco J, Guzmán R, Suter A, Magén C, Morellón L, Ibarra MR, Pardo JA, Algarabel PA (2015) Nature of antiferromagnetic order in epitaxially strained multiferroic $SrMnO_3$ thin films. Phys Rev B 92:024419

20. Moya X, Hueso LE, Maccherozzi F, Tovstolytkin AI, Podyalovskii DI, Ducati C, Phillips LC, Ghidini M, Hovorka O, Berger A et al (2013) Giant and reversible extrinsic magnetocaloric effects in $La_{0.7}Ca_{0.3}MnO_3$ films due to strain. Nat Mater 12(1):52–58

21. Matsunami D, Fujita A, Takenaka K, Kano M (2015) Giant barocaloric effect enhanced by the frustration of the antiferromagnetic phase in Mn_3GaN. Nat Mater 14:73

22. Lukashev P, Sabirianov RF (2010) Spin density in frustrated magnets under mechanical stress: Mn-based antiperovskites. J Appl Phys 107(9):09E115

23. Gama S, Coelho AA, de Campos A, Carvalho AMG, Gandra FCG, von Ranke PJ, de Oliveira NA (2004) Pressure-induced colossal magnetocaloric effect in MnAs. Phys Rev Lett 93:237202

24. Staunton JB, Banerjee R, dos Santos Dias M, Deak A, Szunyogh L (2014) Fluctuating local moments, itinerant electrons, and the magnetocaloric effect: compositional hypersensitivity of FeRh. Phys Rev B 89:054427

25. Sandeman KG (2012) Magnetocaloric materials: The search for new systems. Scr Mater 67(6):566–571. Viewpoint Set No. 51: Magnetic Materials for Energy

26. Zverev VI, Tishin AM, Kuz'min MD (2010) The maximum possible magnetocaloric ΔT effect. J Appl Phys 107(4):043907

27. Moya X, Kar-Narayan S, Mathur ND (2014) Caloric materials near ferroic phase transitions. Nat Mater 13(5):439–50

28. Pecharsky VK, Gschneidner KA Jr (1997) Giant magnetocaloric effect in $Gd_5(Si_2Ge_2)$. Phys Rev Lett 78:4494–4497

Chapter 7
The Magnetism of Mn_3A (A = Pt, Ir, Rh, Sn, Ga, Ge)

The Mn_3A class of magnetic materials, where A = Pt, Ir, Rh, Sn, Ga, and Ge, shows a striking long diversity of different crystal structures, comprising cubic (A = Pt, Ir, Rh), hexagonal (A = Sn, Ga, Ge), and tetragonal (A = Ga, Ge) lattices. These systems are, therefore, challenging tests for the SDFT-DLM theory presented owing to their consequent large variety of magnetic phase transitions and magnetic phases, such as collinear and non-collinear antiferromagnetic states [1–5] and ferrimagnetism [6, 7]. Such a rich magnetic phase complexity has made Mn_3A a very important magnetic materials class with high relevance for spintronics [8–15] and spin chirality exploitation [16]. Furthermore, it also allows to investigate further the fundamental understanding of magnetic frustration produced in three different structures [17, 18], and thus complementing our results obtained in Chap. 6. The present chapter is divided into Sects. 7.1, 7.2.2, and 7.2.3, in which SDFT-DLM theory is applied to the cubic, hexagonal, and tetragonal crystals, respectively. Each of these sections contain a deeper introduction motivating and explaining the corresponding magnetism that will be studied.

7.1 First- and Second-Order Magnetic Phase Transitions in Cubic Mn_3A (A = Pt, Ir, Rh)

7.1.1 Experimental Properties

When A is a transition metal Mn_3A crystallises into the Cu_3Au-type cubic structure, where the Mn atoms are located at the face centres and A atoms occupy the corner positions. Whilst Mn_3Ir and Mn_3Rh show a second-order transition from the PM state to a triangular AFM state (Fig. 7.1a) [19, 20], Mn_3Pt shows a second-order collinear AFM-PM transition when cooling through $T_N = 475\,K$ (Fig. 7.1b) [3–5, 19]. The corresponding magnetic unit cell of the collinear AFM phase is twice (and so tetragonal) the crystallographic unit cell. Sites with non-zero net moment form

© Springer Nature Switzerland AG 2020

E. Mendive Tapia, *Ab initio Theory of Magnetic Ordering*, Springer Theses,
https://doi.org/10.1007/978-3-030-37238-5_7

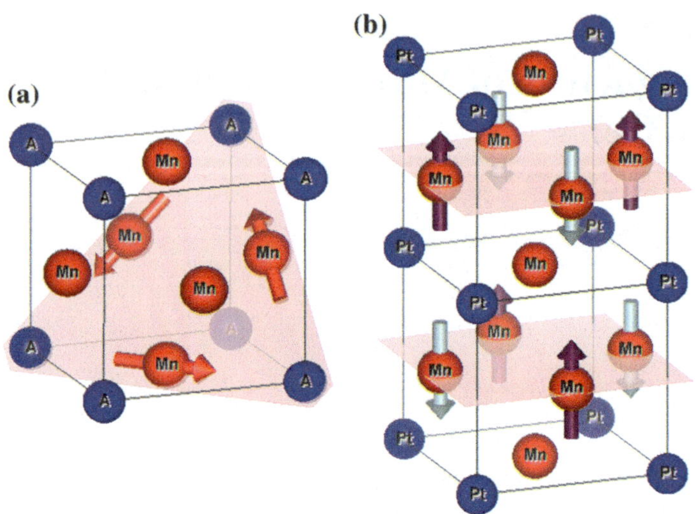

Fig. 7.1 Magnetic unit cells of the **a** triangular AFM and **b** collinear AFM magnetic states in cubic Mn$_3$A. Arrows are used to indicate the net magnetic moment after performing the average over the local moment orientations, i.e. the local order parameters $\{\mathbf{m}_n\}$. Note that for the collinear AFM phase the magnetic sites outside the pink layers have $\mathbf{m}_n = \mathbf{0}$

AFM planes, indicated by pink layers in the figure, stacked perpendicular to the c axis and modulated with a wave vector $\mathbf{q} = (0, 0, 0.5)\frac{2\pi}{a}$, where a is the lattice parameter. The local order parameters describing this state, therefore, completely invert their orientations from one layer to the adjacent ones. The Mn atoms with vanishing net local magnetic order sit in layers staggered between the AFM planes.

Mn$_3$Ir and Mn$_3$Rh do not exhibit another transition and so their triangular AFM state remains stable at lower temperatures. However, reducing the temperature further down to $T_{tr} = 365$ K [4, 5] in Mn$_3$Pt triggers an additional first-order transition from the collinear to the triangular AFM state. An important observation is that the transition temperatures of Mn$_3$Pt change significantly under pressure application with reported values as $dT_N/dp = -70$ K/GPa and $dT_{tr}/dp = 140$ K/GPa [4]. A hydrostatic pressure of $p_c \approx 0.3$ GPa is in consequence enough to completely destroy the collinear AFM order and a unique PM-triangular AFM phase transition is found for $p > p_c$, hence suggesting the presence of a considerable magnetovolume coupling. This is in line with the phenomenological magnetic phase diagram for pairwise interactions constructed by Shirai et al. [21] to investigate the itinerant magnetism of Mn$_3$Pt. In this work the authors suggested that the change of magnetic interactions induced by applied pressures, i.e. a magnetovolume coupling, can produce the magnetic phase transitions observed in Mn$_3$Pt. We will focus now on the study of both high order and magnetovolume coupling effects in order to investigate the origin of this transition.

7.1.2 High Temperature Regime and Magnetovolume Coupling

We firstly inspect the direct correlation function in the PM limit to see if the triangular AFM state for Mn$_3$Rh and Mn$_3$Ir, and the collinear AFM state for Mn$_3$Pt, are the potential stable magnetic phases. Since the crystallographic unit cell contains three Mn atoms, $\tilde{S}^{(2)}_{ss'}(\mathbf{q})$ is a 3×3 matrix with three eigenfunctions $\{\tilde{u}_i(\mathbf{q}), i = 1, 2, 3\}$. We carried out SDFT-DLM calculations for the three magnetic materials at their respective experimental lattice parameters in the PM state. Local magnetic moments with sizes decreasing as $\mu_{\text{Mn(Pt)}} > \mu_{\text{Mn(Rh)}} > \mu_{\text{Mn(Ir)}}$ established at each Mn site. In Table 7.1 we show these values and the experimental lattice parameters used in the calculations.

Figure 7.2a shows the largest eigenvalue of $\tilde{S}^{(2)}_{ss'}(\mathbf{q})$ along the direction (001) in the reciprocal space. We explored the $\mathbf{q}$-dependence of $\tilde{S}^{(2)}_{ss'}(\mathbf{q})$ and verified that there are no other potential magnetic phases, as illustrated for Mn$_3$Pt with $a = 3.95\,\text{Å}$ in Fig. 7.3. Figure 7.2a shows that there are two competing $\mathbf{q}$-points corresponding to $\mathbf{q} = \mathbf{0}$ and $\mathbf{q} = (0, 0, 0.5)\frac{2\pi}{a}$. Pleasingly, we have found that the eigenvector components of $\tilde{S}^{(2)}_{ss'}(\mathbf{q})$ at $\mathbf{q} = \mathbf{0}$ and $\mathbf{q} = (0, 0, 0.5)\frac{2\pi}{a}$ are in direct agreement with the magnetic order found in experiment. For $\mathbf{q} = (0, 0, 0.5)\frac{2\pi}{a}$ they adopt the shape of (++0), i.e. the collinear AFM state that stabilises in Mn$_3$Pt (Fig. 7.1b), and for $\mathbf{q} = \mathbf{0}$ the components are cosines which describe the triangular arrangement that stabilises in the three materials (Fig. 7.1a), i.e. $(1, -\frac{1}{2}, -\frac{1}{2})$. Note that the first two components refer to Mn atoms positioned within the pink layers in Fig. 7.1b, i.e. at (0.5 0 0.5) and (0 0.5 0.5), while the third refers to the site with no net local magnetic moment orientational order, at (0.5 0.5 0), in units of lattice parameters. Remarkably, results for Mn$_3$Pt are in sharp contrast to Mn$_3$Ir and Mn$_3$Rh (Fig. 7.2a). Whilst the peak at $\mathbf{q} = (0, 0, 0.5)\frac{2\pi}{a}$, corresponding to the collinear AFM state, is strongly suppressed for Mn$_3$Ir and Mn$_3$Rh and the triangular structure is therefore the stable state as found in experiment, for Mn$_3$Pt both $\mathbf{q} = \mathbf{0}$ and $\mathbf{q} = (0, 0, 0.5)\frac{2\pi}{a}$ have similar eigenvalues and so similar stability. In Table 7.1 we show the second-order transition temperatures obtained from the largest eigenvalues of $\tilde{S}^{(2)}_{ss'}(\mathbf{q})$, and their comparison with experiment.

Table 7.1 Application of the theory to cubic Mn$_3$A (A = Pt, Ir, Rh). The table shows the lattice parameters used for each material, taken directly from experiment, and theory results for local moment sizes, and Néel transition temperatures, T_N^{theo}, which are compared with experimental values, T_N^{exp}

	a_{exp} (Å)	μ_{Mn} (μ_B)	T_N^{theo} (K)	T_N^{exp} (K)
Mn$_3$Pt [3]	3.87	3.47	≈790	475
Mn$_3$Ir [19]	3.82	2.83	1300	960
Mn$_3$Rh [20]	3.81	3.41	1400	850

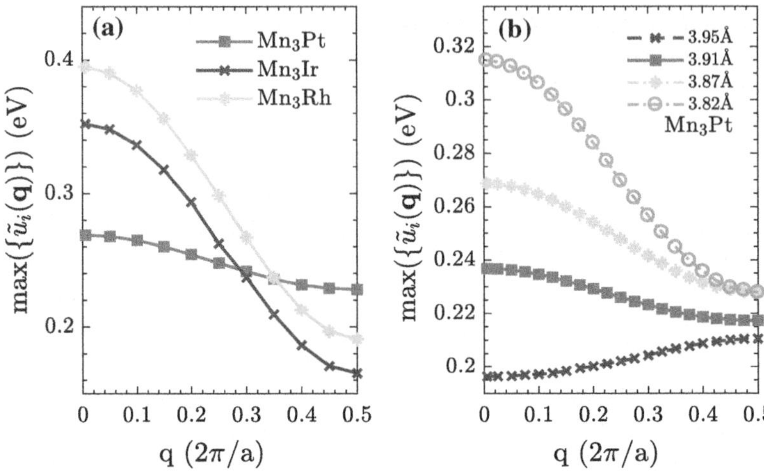

Fig. 7.2 Largest eigenvalue of the direct correlation function along the (001) direction in the reciprocal space for **a** Mn₃Pt (red squares), Mn₃Ir (blue crosses), and Mn₃Rh (turquoise stars) at experimental lattice parameters (see Table 7.1), and **b** for Mn₃Pt for a range of lattice parameters

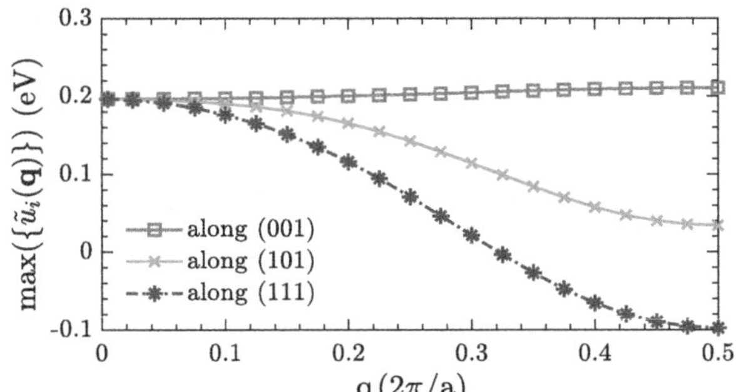

Fig. 7.3 Largest eigenvalue of $\tilde{S}^{(2)}_{ss'}(\mathbf{q})$ of cubic Mn₃Pt versus the wave vector $\mathbf{q}$ for three characteristic directions in the reciprocal space and for a lattice parameter $a = 3.95\,\text{Å}$

Since both triangular AFM at $\mathbf{q} = \mathbf{0}$ and collinear AFM at $\mathbf{q} = (0, 0, 0.5)\frac{2\pi}{a}$ are close in energy in Mn₃Pt, further analysis focused on the magnetovolume coupling and the effect of higher order local moment correlations is required for this material (see Chap. 4). First, we investigate the magnetovolume effect by repeating calculations at different volumes, self-consistently including the effect of local moment magnitude change on the magnetic interactions. We obtained a linear dependence as $\mu_{\text{Mn}} = (3.47 + 3.3\frac{\Delta V}{V_0})\mu_{\text{B}}$ for Mn₃Pt, where V_0 and ΔV are the volume of the unit cell and its relative change.

Figure 7.2b shows that the competition between the collinear and triangular AFM states can be strongly controlled by changing the volume, i.e. the lattice parameter a, indicating the presence of a large magnetovolume effect. For increasing values of a, Mn₃Pt's paramagnetic correlations increasingly favor the collinear AFM state. In fact, for $a = 3.95\,\text{Å}$ the maximum eigenvalue of $\tilde{S}^{(2)}_{ss'}(\mathbf{q})$ peaks at $\mathbf{q} = (0, 0, 0.5)\frac{2\pi}{a}$

and the PM state is, therefore, unstable to the formation of the collinear AFM phase. The absence of this state for A = Ir and Rh evidently is caused by the lattice contraction from the presence of these transition metals. Note that the experimental lattice parameters of these two materials are significantly contracted compared to Mn$_3$Pt. The paramagnetic correlations in Mn$_3$Ir and Mn$_3$Rh thus strongly favor the triangular AFM state. This is further confirmed from the results shown in Sect. 7.1.3, where we calculate the contribution from higher order correlations and complete our finite temperature study of Mn$_3$Pt. After minimising the free energy we obtain a critical value of the lattice parameter a_c such that structures with $a < a_c$ show a single PM-triangular AFM second-order phase transition for Mn$_3$Pt. We find that the collinear AFM state is not stable in Mn$_3$Ir and Mn$_3$Rh even for expansions around a_c.

7.1.3 Lower Temperature Regime and the Magnetic Phase Diagram of Mn$_3$Pt

We now use the SDFT-DLM theory to calculate the internal magnetic fields $\{\mathbf{h}_n^{\text{int}}\}$ as functions of the local order parameter, extract the higher order local moment interactions and produce the free energy. Here the triangular AFM phase is formed by local order parameters forming $120°$, and so a single size m_{tri} associated with each of the three Mn sites describes this magnetic state. Similarly, by symmetry the magnitudes among $\{\mathbf{m}_n\}$ at sites with non-zero net magnetic moment are the same for the collinear state, that we label as m_{coll}. They have opposite directions as shown in Fig. 7.1b. We define h_{tri}^{int} and h_{coll}^{int} as the absolute values of the effective fields sustaining the local moments for the triangular and collinear AFM states, respectively. In Fig. 7.4 we show their dependence on m_{tri} and m_{coll}. Importantly, while h_{coll}^{int} exhibits a linear dependence, h_{tri}^{int} shows a more complicated behaviour with a negative effect from higher than linear order coefficients. From this fact it directly follows that high order local moment interactions, $\mathcal{S}_{ijkl}^{(4)}$, cannot stabilize the triangular state and trigger the first-order AFM-AFM transition at lower temperature, which reinforces the idea that the magnetovolume coupling is the dominant factor of this transition.

The free energy $\mathcal{G}_1$ per formula unit (three Mn's and one Pt atoms) accounting for the effect of a hydrostatic pressure p can be written using Eqs. (3.20) and (4.1)

$$\mathcal{G}_1 = \Omega_0 + f^{(2)}\left(\mathbf{m}_1, \mathbf{m}_2, \mathbf{m}_3; \omega\right) + f^{(4)}\left(\mathbf{m}_1, \mathbf{m}_2, \mathbf{m}_3; \omega\right)$$
$$+ \frac{1}{2}V_0\gamma\omega^2 + p\omega V_0 - T\left(S_1(\mathbf{m}_1) + S_2(\mathbf{m}_2) + S_3(\mathbf{m}_3)\right), \tag{7.1}$$

where pertinent terms describing an elastic energy and the cost of p have been added, with compressibility γ being taken from experiment with a value 111 GPa [4]. Note that the local moment correlation functions, $f^{(2)}$ and $f^{(4)}$, have an explicit dependence on ω, which describes the magnetovolume coupling. We follow the

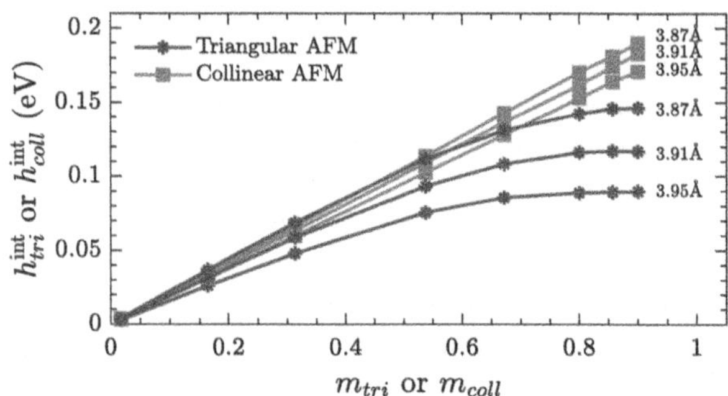

Fig. 7.4 Absolute value of the internal fields, h_{tri}^{int} and h_{coll}^{int}, for the triangular (red squares) and collinear (blue stars) AFM states as functions of magnetic order and for a range of lattice parameters

procedure explained in Chap. 4 and fit the internal fields $h_{tri}^{int}(m_{tri})$ and $h_{coll}^{int}(m_{coll})$ for different magnetic orderings, as well as different lattice parameters to extract their dependence on volume. We find that the 8 data points in Fig. 7.4 for each curve can be very well fitted by

$$
\begin{aligned}
h_{tri}^{int} &= \mathcal{S}_{tri}^{(2)}(\omega)m_{tri} + \mathcal{S}_{tri}^{(4)}m_{tri}^3, \\
h_{coll}^{int} &= \mathcal{S}_{coll}^{(2)}(\omega)m_{coll},
\end{aligned}
\tag{7.2}
$$

where $\mathcal{S}_{tri/coll}^{(2)}(\omega)$ and $\mathcal{S}_{tri/coll}^{(4)}$ are compact forms of pair and quartet local moment interactions for the magnetic phases under study. The fitting of Eq. (7.2) shows that $\mathcal{S}_{tri}^{(4)} \neq 0$, $\mathcal{S}_{coll}^{(4)} = 0$, and that higher order terms are negligible. Both collinear and triangular states have been found to follow a good linear dependence on ω for the pairwise contribution only, $\mathcal{S}_{coll}^{(2)} = (\mathcal{S}_{coll,0}^{(2)} + \alpha_{coll}\omega) = (202 - 324\omega)\,\mathrm{meV}$ for the collinear state, and $\mathcal{S}_{tri}^{(2)}(\omega) = (\mathcal{S}_{tri,0}^{(2)} + \alpha_{tri}\omega) = (192 - 1086\omega)\,\mathrm{meV}$ and $\mathcal{S}_{tri}^{(4)} = -80\,\mathrm{meV}$ for the triangular state. From Eqs. (4.1)–(4.4) the internal energy becomes

$$
\begin{aligned}
\langle\Omega^{int}\rangle_0\Big|_{\{\mathbf{m}_i\}_{tri}} &= \Omega_0 - \frac{3}{2}\mathcal{S}_{tri}^{(2)}(\omega)m_{tri}^2 - \frac{3}{4}\mathcal{S}_{tri}^{(4)}m_{tri}^4, \\
\langle\Omega^{int}\rangle_0\Big|_{\{\mathbf{m}_i\}_{coll}} &= \Omega_0 - \mathcal{S}_{coll}^{(2)}(\omega)m_{coll}^2,
\end{aligned}
\tag{7.3}
$$

where $\{\mathbf{m}_i\}_{tri/coll}$ means that the local order parameters are set to describe the triangular/collinear AFM states. We stress again that whilst the triangular state has three non-zero local order parameters inside the cell, the collinear state has only two. This is due to the site with zero net magnetisation, which we refer to as the third magnetically disordered sublattice, i.e. $m_3 = 0$ for the collinear AFM state. We now minimise $\mathcal{G}_1$ with respect to ω analytically and write from Eqs. (7.1) and (7.3) that

$$
\begin{aligned}
\omega\Big|_{\{\mathbf{m}_i\}_{tri}} &= \frac{\alpha_{tri}}{V_0\gamma}\frac{3}{2}m_{tri}^2 - \frac{p}{\gamma}, \\
\omega\Big|_{\{\mathbf{m}_i\}_{coll}} &= \frac{\alpha_{coll}}{V_0\gamma}m_{coll}^2 - \frac{p}{\gamma},
\end{aligned}
\tag{7.4}
$$

where $\alpha_{tri} = \sum_{ij} \alpha_{tri,ij}$ and $\alpha_{coll} = \sum_{ij} \alpha_{coll,ij}$. Substituting Eq. (7.4) into Eq. (7.1) gives

$$
\begin{aligned}
\mathcal{G}_1 \Big|_{\{\mathbf{m}_i\}_{tri}} =&\, \Omega_0 - 3 \left[\frac{1}{2} \mathcal{S}_{tri,0}^{(2)} - \frac{p\alpha_{tri}}{\gamma} \right] m_{tri}^2 - \frac{3}{4} \left[\mathcal{S}_{tri}^{(4)} + \frac{3\alpha_{tri}^2}{2V_0\gamma} \right] m_{tri}^4 \\
& - T \Big(S_1(m_{tri}) + S_2(m_{tri}) + S_3(m_{tri}) \Big) - p^2 V_0 \gamma^{-1}, \\
\mathcal{G}_1 \Big|_{\{\mathbf{m}_i\}_{coll}} =&\, \Omega_0 - 2 \left[\frac{1}{2} \mathcal{S}_{coll,0}^{(2)} - \frac{p\alpha_{coll}}{\gamma} \right] m_{coll}^2 - \frac{1}{2} \frac{\alpha_{coll}^2}{V_0\gamma} m_{coll}^4 \\
& - T \Big(S_1(m_{coll}) + S_2(m_{coll}) + S_3(0) \Big) - p^2 V_0 \gamma^{-1}.
\end{aligned}
\tag{7.5}
$$

Hence, the magnetovolume coefficients $\{\alpha_{tri}, \alpha_{coll}\}$ give rise to fourth order contributions that in principle can trigger the first-order AFM-AFM phase transition. This is indeed what we have found in our calculations after minimising Eq. (7.5) at $p = 0$ at different temperatures. We find the first-order AFM-AFM transition to occur for lattice parameters $a > a_c = 3.93\,\text{Å}$, from which both transition temperatures exist with $T_N > T_{tr}$. Below this value the collinear AFM state is entirely suppressed and a single second-order triangular AFM-PM transition occurs. $T_N \approx 800\,\text{K}$ for $a > a_c$, somewhat higher than the experimental value of 475 K [3–5]. Notably, our DFT-DLM approach correctly predicts the two triangular and collinear AFM orderings as the most stable phases and explains the occurrence of the first-order transition as a magnetovolume driven effect.

To construct the pressure-temperature magnetic phase diagram we firstly choose a reference lattice parameter for the PM state that best describes the $p = 0$ state, i.e. $a_{\text{PM}} = 3.98\,\text{Å}$ ($V_0 = a_{\text{PM}}^3$). This sufficiently expands the unit cell to stabilise the collinear AFM state at high T and gives a temperature span for its stability of $T_N - T_{tri} = 125\,\text{K}$, which agrees well with the experimental value after rescaling with our higher T_N. We show in Fig. 7.5 the ab initio magnetic phase diagram obtained. For increasing values of p, T_{tr} rises and eventually reaches a tricritical point (A) at $p_c = 0.33\,\text{GPa}$, remarkably close to the experimental value [4]. At higher values $p > p_c$ the collinear AFM order disappears at all temperatures. Our method is able to distinguish between second- and first-order phase transitions, which we indicate with continuous and dashed lines in the figure. Moreover, the volume change at T_{tr} and $p = 0$ is $\omega = 0.6\%$, significant but somewhat below experimental findings (2.4%).

7.1.4   On the Deficiencies of the Collinear AFM State

To describe vanishing magnetic order from fluctuating local moments on sublattices at the third sites in the collinear AFM phase, we have kept $m_3 = 0$ even at temperatures well below T_N. However, in the limit of approaching $T = 0\,\text{K}$ all magnetic sites should be fully ordered at every magnetic phase, i.e. $\{m_n = 1\}$. This means that some

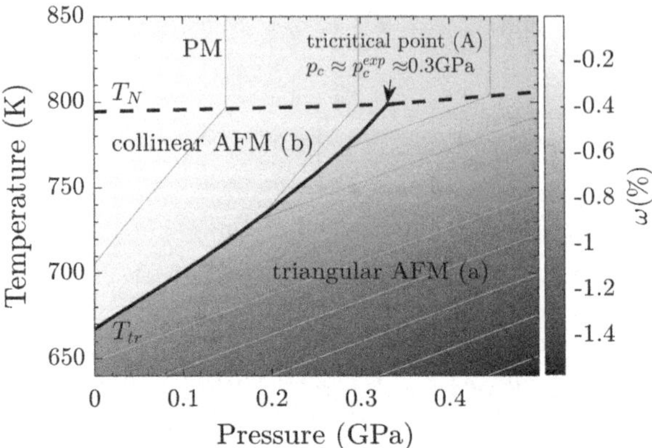

Fig. 7.5 Ab initio pressure-temperature magnetic phase diagram of Mn₃Pt. The colour scheme encodes the relative change of volume with respect to the PM state, ω. Thick black lines indicate first-order (solid) and second-order (dashed) magnetic phase transitions and letters in brackets link to panels of Fig. 7.1. A tricritical point (A) with $p_c \approx p_c^{exp}$ [4] is marked

net magnetic order should develop at the third site at intermediate temperatures, even if the triangular state was not stabilised. This is something that was pointed out by Long [17], who noted the deficiencies of such a collinear AFM state. He thus proposed a new magnetic structure compatible with a modulation of $\mathbf{q} = (0, 0, 0.5)\frac{2\pi}{a}$ and with non-zero magnetic moment densities at every magnetic site. To study this effect we have calculated the local magnetic fields for magnetic phases in which only the third magnetic site has non-zero magnetic ordering, $\{m_1 = m_2 = 0, m_3 \neq 0\}$, in both FM ($\mathbf{q} = \mathbf{0}$) and collinear AFM ($\mathbf{q} = (0, 0, 0.5)\frac{2\pi}{a}$) modulations. From this calculation we can investigate at what temperature this site begins to develop net spin polarisation when embedded in an otherwise PM environment, i.e. we study the effect from the leading second-order coefficients in the free energy. In both cases we find that the electronic interactions set a non-zero magnetic order at the third site only for temperatures around 100 K or below. These values are very small compared with the transition temperatures and so strongly indicate that the collinear AFM state as described in Fig. 7.1b should persist down to T_{tr}. In addition, we also show that the magnetovolume coupling produces the leading quartic coefficients in the interacting part of free energy $\mathcal{G}_1$. Thus, as the temperature decreases the triangular AFM phase stabilises before $m_3 \neq 0$ occurs for the collinear AFM state. As suggested by Hirano et al. [22], however, further measurements are necessary to corroborate the nature of the collinear AFM phase. We finally remark that our analysis adds to the findings from Kota et al. DFT calculations [23].

7.1.5 The Barocaloric Effect of Mn$_3$Pt

We finish the investigation of Mn$_3$Pt by evaluating its BCE. We follow a similar procedure as performed in Chap. 6. After inspecting Fig. 7.5 clearly the largest thermodynamic changes occur around the first-order AFM-AFM phase transition T_{tr} at $p = 0$. For increasing values of p the transition loses its first-order character and so the associated entropy change becomes smaller. We calculate the total entropy difference above and below T_{tr} using Eq. (4.28), as well as Eqs. (3.24) and (4.25). For the corresponding phases the local order parameters are $m_{coll} = 0.495$ and $m_{tri} = 0.431$, respectively. From this it follows that the magnetic contribution is small, $\Delta S_{\mathrm{mag}} = 2\,\mathrm{J/kgK}$. Moreover, ΔS_{elec} is negligible owing to the tiny change of the density of states at the Fermi energy between these two phases. Despite adiabatic temperature changes being potentially large due to the large value of dT_{tr}/dp, the isothermal entropy change is small meaning that Mn$_3$Pt is not a good candidate for cooling based on hydrostatic pressure application, unlike the Mn-based antiperovskites [24, 25].

7.2 Mn$_3$A (A = Sn, Ga, Ge): Hexagonal and Tetragonal Structures

7.2.1 Experimentally Observed Properties and Published Theoretical Calculations

At high temperatures Mn$_3$A (A = Sn, Ga, Ge) can crystallise into the hexagonal DO_{19}-type structure shown in Fig. 7.6 [2, 6, 7, 9, 26–28]. Equidistant layers containing both Mn and A atoms, preserving the formula unit proportion, are stacked perpendicular to the c axis. Owing to the presence of geometrically frustrated AFM interactions, a triangular AFM order, as illustrated by the arrows in the figure, forms in experiment below a Néel temperature that spans around 400 K depending on the choice of A, as indicated in Table 7.2 [6, 7, 9, 26–28]. Much interest for these systems is driven by the fact that a strong anomalous Hall effect (AHE) has been experimentally observed for all three materials (A = Sn, Ga, Ge) [29–31], as well as by the demonstration from purely symmetry arguments that spin-polarized currents can be induced owing to the non-collinearity [15]. Contrary to chiral FM structures with a non-zero out-of-plane component, in which the AHE can be driven by a real-space topological effect [32, 33], in co-planar chiral antiferromagnets with local magnetic moments lying within the basal plane the AHE is generated by the spin-orbit coupling (SOC) [34]. The hexagonal family Mn$_3$A shows a SOC that distorts the triangular arrangement, generating a weak non-zero net magnetisation [1, 2, 27, 35]. These systems have been in consequence focus of extensive theoretical work based on $T = 0$ K ab initio calculations in order to evaluate their potential to induce large Hall effects and study possible triangular configurations [8, 16, 34, 36].

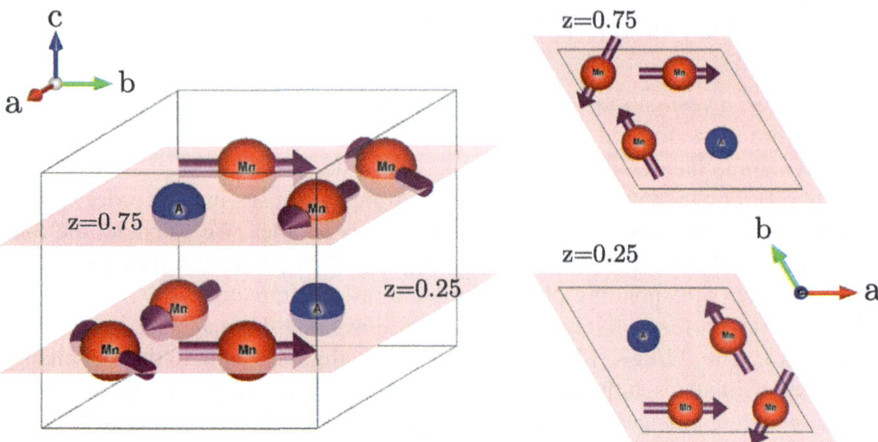

Fig. 7.6 Hexagonal magnetic structure of Mn$_3$A (A = Sn, Ga, Ge). Magenta arrows are along the local magnetic order parameters $\{\mathbf{m}_n\}$, and z indicates the height with respect to the $\hat{c}$ lattice direction. The figure shows the structure for the ideal case of a fully compensated triangular state

Moreover, on annealing for long periods of time the hexagonal Mn$_3$Ga and Mn$_3$Ge phases, at low temperature they transform into the tetragonal DO_{22} phase shown in Fig. 7.9a [6, 7, 37, 38]. This structure contains two non-equivalent atomic positions denoted as $2b$ and $2d$. The magnetic phase that stabilises is in sharp contrast with the triangular AFM state observed in its hexagonal counterpart. It is a ferrimagnetic (FIM) state in which in a unit cell two of the Mn magnetic local order parameters point in one direction and the remaining along the opposite. These directions are indicated by magenta and blue arrows in the figure and are associated with the $4d$ and $2b$ positions, respectively. The magnetism here is collinear so that spintronic properties from non-collinear and chiral AFM are not present [16]. However, the FIM phase has a high Curie temperature and a low net magnetic moment. This together with the fact that epitaxial films can be grown to achieve high perpendicular magnetic anisotropy have made these systems very appealing for spin transfer torque applications set in small and thermally stable designs [12, 13, 39–43].

A consistent temperature-dependent study of hexagonal and tetragonal Mn$_3$A, although highly desirable, is still missing. In this section we address this by applying our DLM picture without SOC and so study the consequences of thermal fluctuations on magnetic frustration and magnetic phase stability. The results obtained can be used as a starting point for subsequent studies which include relativistic effects and which are pitched at uncovering subtle SOC effects on electronic and magnetic structure at finite temperatures.

Table 7.2 Application of SDFT-DLM theory to hexagonal Mn$_3$A (A = Sn, Ga, Ge). The table shows the lattice parameters taken directly from experiment, and results for local moment sizes, direct pair $\mathcal{S}^{(2)}_{tri}$ and higher order local moment correlations, $\mathcal{S}^{(4)}_{tri}$ and $\mathcal{S}^{(6)}_{tri}$, and Néel transition temperatures, T_N^{theo}. The latter are compared with experimental values, T_N^{exp}

Reference	Mn$_3$Sn [26, 27]	Mn$_3$Ga [6]	Mn$_3$Ge [7, 9, 28]
$a = b$ (Å)	5.66	5.40	5.36
c (Å)	4.53	4.35	4.32
T_N^{exp} (K)	420	470	365
T_N^{theo} (K)	495	586	486
μ_{Mn} (μ_B)	2.98	2.65	2.43
$\mathcal{S}^{(2)}_{tri}$ (meV)	131	152	127
$\mathcal{S}^{(4)}_{tri}$ (meV)	−80	−61.5	−68
$\mathcal{S}^{(6)}_{tri}$ (meV)	59	19.5	46

7.2.2 Application of DLM Theory to Hexagonal Mn$_3$Sn, Mn$_3$Ga and Mn$_3$Ge

The first step is to calculate the direct correlation function $\tilde{\mathcal{S}}^{(2)}_{ss'}(\mathbf{q})$ and obtain the most stable magnetic phases. For the three hexagonal materials Mn$_3$Sn, Mn$_3$Ga and Mn$_3$Ge under study local magnetic moments with similar sizes established at each Mn site. In Table 7.2 we give these values as well as the lattice parameters used, taken from experiment. Since the crystallographic unit cell of the hexagonal Mn$_3$A contains six magnetic positions, $\tilde{\mathcal{S}}^{(2)}_{ss'}(\mathbf{q})$ is a 6×6 matrix. Figure 7.7a–c shows the largest eigenvalue of $\tilde{\mathcal{S}}^{(2)}_{ss'}(\mathbf{q})$, $\max\{\tilde{u}_a(\mathbf{q})\}$, corresponding to the most stable magnetic phase, against the wave vector inside the Brillouin zone. For the three materials it peaks at $\mathbf{q} = \mathbf{0}$. This means that the magnetic unit cell matches the crystallographic unit cell and so there are no rotations of the local moment directions from one cell to another. An inspection of the eigenvector components gives the relative orientations between the six local order parameters inside one magnetic unit cell. For the three systems these components perfectly match cosines of angles describing a fully compensated triangular state as shown in Fig. 7.6. The eigenvector is for example $(1, -\frac{1}{2}, -\frac{1}{2}, 1, -\frac{1}{2}, -\frac{1}{2})$, where the first three components correspond to Mn atoms at z = 0.25 and the next three at z = 0.75. After examining the $\mathbf{q}$-dependence of $\tilde{\mathcal{S}}^{(2)}_{ss'}(\mathbf{q})$ thoroughly we did not find signatures of $\mathbf{q} \neq \mathbf{0}$ phases that would be more stable, i.e. the magnetic correlations that are precursors of the triangular state are the strongest. Importantly, in striking contrast to cubic Mn$_3$A (A = Pt, Ir, Rh) here we found that the eigenvalue of $\tilde{\mathcal{S}}^{(2)}_{ss'}(\mathbf{q})$ at $\mathbf{q} = \mathbf{0}$ is consistently the largest for different unit cell volumes, and therefore magnetoelastic effects are not relevant.

We now model the triangular AFM state at lower temperatures. This firstly requires that this magnetic state is described in terms of the local order parameters $\{\mathbf{m}_n\}$, which describe the magnetic order of the triangular AFM phase in Fig. 7.6. Since this state

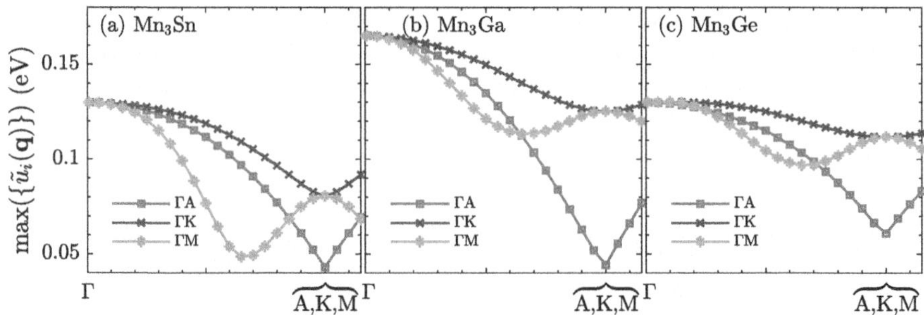

Fig. 7.7 Dependence of the largest eigenvalue of $\tilde{S}^{(2)}_{ss'}(\mathbf{q})$ for hexagonal **a** Mn₃Sn, **b** Mn₃Ge, and **c** Mn₃Ga. The results are given for three characteristic directions within the Brillouin zone through the Γ point

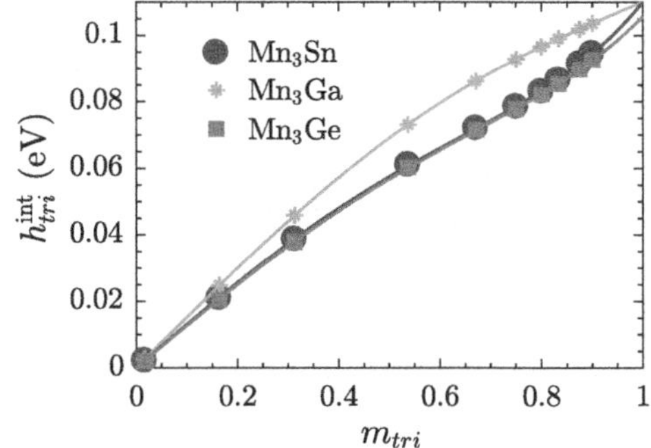

Fig. 7.8 Strength of the local field against the local order parameter for the three studied materials, Mn₃Sn (black circles), Mn₃Ga (light blue stars), and Mn₃Ge (magenta squares). Continuous lines are the fitted functions of Eq. (7.6) using the data points

is fully compensated, the six vectors $\{\mathbf{m}_n\}$ in the unit cell have the same length and only differentiate by in-plane rotations of 120°. Lowering T does not change their orientations and induces an identical increase of their magnitudes only. Hence, the entire magnetic state is characterised by one common local order parameter size that we label as m_{tri}. Evidently, the effective fields $\{\mathbf{h}_n^{\text{int}}\}$, given in Eq. (3.28), have the same length h_{tri}^{int} at each Mn site too. In Fig. 7.8 we show the DFT-DLM calculation of h_{tri}^{int} as a function of m_{tri} for A = Sn, Ga, and Ge. Although small, there is a clear deviation from a linear dependence for increasing m_{tri}, which explicitly indicates that higher order local moment interaction effects are significant (see Sect. 4.1). Indeed, we have found that the local field data is very well fitted for every choice of atom A by

$$h_{tri}^{\text{int}} = \mathcal{S}^{(2)}_{tri} m_{tri} + \mathcal{S}^{(4)}_{tri} m_{tri}^3 + \mathcal{S}^{(6)}_{tri} m_{tri}^5, \tag{7.6}$$

where $\mathcal{S}^{(2)}_{tri}$ is directly determined from the largest eigenvalue of $\tilde{S}^{(2)}_{ss'}(\mathbf{q})$, which corresponds to the triangular AFM state, as $\mathcal{S}^{(2)}_{tri} = \max\{\tilde{u}_a(\mathbf{q} = \mathbf{0})\}$. We remark that we have used 10 data points, shown as single points in Fig. 7.8, to fit the two constants $\mathcal{S}^{(4)}_{tri}$ and $\mathcal{S}^{(6)}_{tri}$. Their values are tabulated in Table 7.2, and the fitted functions are shown in Fig. 7.8 as continuous lines.

As explained in Sect. 4.1, $\mathcal{S}_{tri}^{(2)}$ and $\{\mathcal{S}_{tri}^{(4)}, \mathcal{S}_{tri}^{(6)}\}$ are compact forms of local moment interactions for the magnetic state under study, which is the triangular AFM state in this case. $\mathcal{S}_{tri}^{(4)}$ and $\mathcal{S}_{tri}^{(6)}$ describe the feedback between the spin-polarised electronic structure and the increase of magnetic order. Their calculation from the data of Fig. 7.8 directly provides the magnetic material's free energy from Eqs. (3.20), (3.28), and (4.1),

$$\mathcal{G}_1 = \Omega_0 - \frac{1}{2}\mathcal{S}_{tri}^{(2)} m_{tri}^2 - \frac{1}{4}\mathcal{S}_{tri}^{(4)} m_{tri}^4 - \frac{1}{6}\mathcal{S}_{tri}^{(6)} m_{tri}^6 - T S_n(m_{tri}), \qquad (7.7)$$

per magnetic site n. Minimising Eq. (7.7) with respect to m_{tri} yields m_{tri} itself as a function of T. The values of $\mathcal{S}_{tri}^{(4)}$ and $\mathcal{S}_{tri}^{(6)}$ obtained are negative or small so that the triangular AFM-PM transition at T_N^{theo} has been found to be of second-order, in agreement with experiment. Remarkably, the theory correctly captures trends and magnitudes of T_N for different choices of atom A in line with experimental findings as shown in Table 7.2. Interestingly, from Fig. 7.8d we observe that Mn$_3$Sn and Mn$_3$Ge have a similar behaviour in qualitatively significant contrast compared with Mn$_3$Ga. The main differences are the higher gradient of h_{tri}^{int} at small values of m_{tri} for Mn$_3$Ga, leading to the larger values of both T_N^{theo} and $\mathcal{S}_{tri}^{(2)}$, as well as the different concavity around $m_{tri} \approx 0.8$ as a direct consequence of the presence of $\mathcal{S}_{tri}^{(4)}$ and $\mathcal{S}_{tri}^{(6)}$. These higher order components are significant in hexagonal Mn$_3$A, and with the consequence that any effective pairwise interactions between local moments calculated for different magnetic states will adopt different values.

7.2.3   Application of DLM Theory to Tetragonal Mn$_3$Ga and Mn$_3$Ge

The tetragonal structure shown in Fig. 7.9a can be equivalently described as a bcc lattice with four atoms per unit cell. One Ga atom is positioned at $(0, 0, 0)$ and three Mn atoms at $\mathbf{r}_{\text{Mn}_1} = (0, 0, 1/2)$, $\mathbf{r}_{\text{Mn}_2} = (1/2, 0, 1/4)$, and $\mathbf{r}_{\text{Mn}_3} = (0, 1/2, 1/4)$. The last two atomic positions are equivalent and associated with the $4d$ site, while the first Mn position corresponds to the $2b$ site. Similarly to the Cu$_3$Au-type case in Mn$_3$Pt, the direct correlation function $\tilde{\mathcal{S}}_{ss'}^{(2)}(\mathbf{q})$ in the bcc basis is a 3×3 matrix with three eigenfunctions $\{\tilde{u}_i(\mathbf{q}), i = 1, 2, 3\}$. In Fig. 7.9b, c we show the largest value among these functions for relevant directions inside the Brillouin zone. We remark that we consistently obtained magnetic moment sizes $\mu_1 > \mu_2 = \mu_3$, as observed experimentally [6, 39] and in agreement with other first principles calculations [14]. From Fig. 7.9b–c we can see that the peak is at $\mathbf{q} = \mathbf{0}$ and that there are no other competing modulations. From Eq. (3.76) we calculate the FIM-PM transition temperature to be $T_{tr}^{\text{Ga}} = 995\,\text{K}$ and $T_{tr}^{\text{Ge}} = 814\,\text{K}$. Since in experiment the material transforms into the hexagonal structure with increase in temperature before going through T_{tr}, our results cannot be directly compared. Our data is in agreement with literature, however, because the structural transition occurs below our transition temperature predictions for Mn$_3$Ga [6] and the extrapolated value for Mn$_3$Ge, 920 K [39].

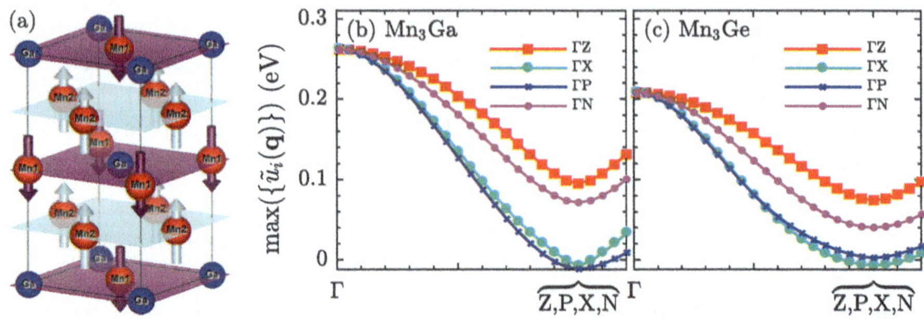

Fig. 7.9 a Ferrimagnetic unit cell of tetragonal Mn$_3$Ga and Mn$_3$Ge. Magenta and blue arrows are used to indicate "up" and "down" local order parameter orientations. **b–c** Wave vector dependence of the largest eigenvalue of $\tilde{\mathcal{S}}_{ss'}^{(2)}(\mathbf{q})$ for Mn$_3$Ga and Mn$_3$Ge. Results are shown for four characteristic directions from the Γ point to special points on the Brillouin zone boundary

Fig. 7.10 Absolute values of the local fields for the two non-equivalent sites, associated with Mn$_1$ and Mn$_2$(Mn$_3$), against the local order parameters when $\mathbf{m} = \mathbf{m}_1 = \mathbf{m}_2 = -\mathbf{m}_3$. Lighter and wider lines are used for Mn$_3$Ga and Mn$_3$Ge, respectively

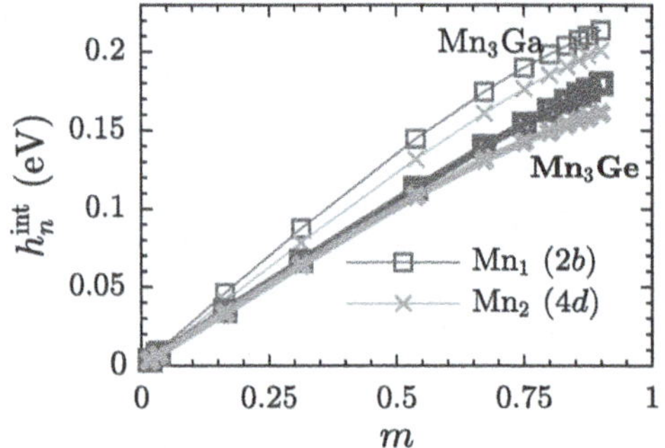

The eigenvector components obtained for the maximum eigenvalue after diagonalising $\tilde{S}_{ss'}^{(2)}(\mathbf{0})$ are $(-0.614, +0.558, +0.558)$ and $(-0.587, +0.573, +0.573)$, for Mn$_3$Ga and Mn$_3$Ge respectively. These describe the FIM phase with Mn atoms at $4d$ sites developing identical local order parameters being antiparallel to the atom at position $2b$. The different values among the eigenvector components indicate that Mn$_1$ develops an average local magnetisation higher than Mn$_2$ and Mn$_3$, i.e. $m_1 > m_2 = m_3$. This becomes more evident after studying how the local fields behave if identical amount of local magnetic ordering is imposed at all sites, which we prescribe by a quantity labelled as m, that is $\mathbf{m} = \mathbf{m}_2 = \mathbf{m}_3 = -\mathbf{m}_3$. Indeed, Fig. 7.10 shows that $h_n^{\text{int}}(m)$ at Mn$_1$ is slightly larger compared with Mn$_{2(3)}$.

We fit the local field DFT-DLM data, $\{\mathbf{h}_n^{\text{int}}\}$, in order to extract the first derivatives of the internal magnetic energy $\langle \Omega^{\text{int}} \rangle_0$ with respect to the local order parameters via Eq. (3.28). We have used the data points shown in Fig. 7.10 as well as many other magnetic configurations comprising $\{\mathbf{m}_1 = -m_1\hat{z}, \mathbf{m}_2 = \mathbf{m}_3 = \mathbf{0}\}$, $\{\mathbf{m}_1 = \mathbf{0}, \mathbf{m}_2 = \mathbf{m}_3 = m_2\hat{z}\}$, and $\{\mathbf{m}_1 = -m_1\hat{z}, \mathbf{m}_2 = \mathbf{m}_3 = m_2\hat{z}\}$, for mixed magnetic orderings ranging as m_1(and/or m_2) = 0.05, 0.1, 0.5, 1, 2, 3, 4, 5, 6, 7, 8, and 10, which thoroughly samples the $\{\mathbf{m}_n\}$-space for the FIM state. In total there were 216 data points to fit, which we found to be very well described by

$$\langle\Omega^{\text{int}}\rangle_0 = -\frac{1}{2}\mathcal{S}_1^{(2)}m_1^2 - \frac{1}{2}\mathcal{S}_2^{(2)}(m_2^2 + m_3^2) - \mathcal{S}_3^{(2)}\mathbf{m}_1 \cdot (\mathbf{m}_2 + \mathbf{m}_3) - \frac{1}{4}\mathcal{S}_1^{(4)}(m_2^4 + m_3^4)$$
$$- \mathcal{S}_1^{(6)}\mathbf{m}_1^3 \cdot (\mathbf{m}_2^3 + \mathbf{m}_3^3) - \mathcal{S}_2^{(6)}m_1^2(m_2^4 + m_3^4)$$
$$- \mathcal{S}_3^{(6)}\mathbf{m}_1 \cdot (\mathbf{m}_2^5 + \mathbf{m}_3^5) - \mathcal{S}_4^{(6)}\mathbf{m}_1^5 \cdot (\mathbf{m}_2 + \mathbf{m}_3).$$

$$(7.8)$$

The higher order terms, compactly contained in $\{\mathcal{S}_1^{(4)}, \mathcal{S}_1^{(6)}, \mathcal{S}_2^{(6)}, \mathcal{S}_3^{(6)}, \mathcal{S}_4^{(6)}\}$, are only important to numerically capture small deviations of a nearly linear dependence of $\{h_n^{\text{int}}\}$, as it can be seen from Fig. 7.10. The multi-spin, or multi-site, local moment interactions that are inferred from these quantities for tetragonal Mn$_3$A, as also proposed by Khmelevskyi et al. [18], have only a small effect on the FIM phase and in consequence the FIM-PM phase transition is predicted to be of second-order. We show the results in Table 7.3. The free energy expressions and Eq. (7.8) for both Mn$_3$Ga and Mn$_3$Ge are identical in form and the sizes of constants are also similar. In particular $\mathcal{S}_1^{(2)}$ is very small compared with $\mathcal{S}_2^{(2)}$ and $\mathcal{S}_3^{(2)}$. This means that the net magnetic polarisation at site 2b arises as a slave effect from the stronger and leading local magnetic ordering at sites 4d. This is underpinned by the large value of $\mathcal{S}_3^{(2)}$, whose negative sign captures the AFM nature among interactions between both sites. We also remark that the higher transition temperature for A = Ga is a direct consequence of the larger values of $\mathcal{S}_2^{(2)}$ and $\mathcal{S}_3^{(2)}$, which links to the temperature from which 4d begins to develop a net local magnetic order.

We finish this section with a discussion of geometrically frustrated magnetism. To this end we analysed in more detail the direct local moment-local moment cor-

Table 7.3 Quadratic and higher order terms as given in Eq. (7.8) for the FIM state of tetragonal Mn$_3$Ga and Mn$_3$Ge. The lattice parameters are taken from experiment [6, 7, 37, 39]. The table also shows the magnetic moment sizes and transition temperatures obtained

	Mn$_3$Ga	Mn$_3$Ge
a (Å)	3.91	3.81
c (Å)	7.10	7.26
μ_1	2.32 μ_B	2.19 μ_B
$\mu_2 = \mu_3$	2.02 μ_B	1.56 μ_B
T_{tr}^A (K)	995	814
$\mathcal{S}_1^{(2)}$	9.60	12.7
$\mathcal{S}_2^{(2)}$	118	110
$\mathcal{S}_3^{(2)}$	−131.5	−99.56
$\mathcal{S}_1^{(4)}$	≈0	−27.6
$\mathcal{S}_1^{(6)}$	54	36.3
$\mathcal{S}_2^{(6)}$	47	35.6
$\mathcal{S}_3^{(6)}$	16	11
$\mathcal{S}_4^{(6)}$	−10.8	−7.9

relation function $\tilde{S}_{ss'}^{(2)}(\mathbf{q})$. We found that parallel alignment of $\mathbf{m}_2$ and $\mathbf{m}_3$ is energetically preferable and compatible with a picture of magnetic interactions that do not show frustration effects. We therefore concluded that tetragonal Mn$_3$A should not be regarded as a magnetically frustrated system. Our findings are in agreement with a major theoretical work on pairwise interactions carried out by Khmelevskyi et al. [18].

7.3   Summary and Conclusions

We have studied magnetic materials in the family Mn$_3$A, currently attracting much interest for spintronic applications. Their rich spectrum of magnetic phases, including non-collinear and collinear AFM and ferrimagnetic states, sets a challenging test for our ab initio theory to explain their temperature-dependent properties and magnetic frustration. For all cubic, hexagonal and tetragonal crystal structures we firstly calculated $\tilde{S}_{ss'}^{(2)}(\mathbf{q})$. Our calculations are in agreement with experiment: we predict the stabilisation of the triangular AFM state for the cubic and hexagonal lattices, the ferrimagnetic state for the tetragonal lattice, as well as a collinear AFM phase for the cubic Mn$_3$Pt, which competes in stabilisation with the triangular AFM state. The values and trends of transition temperatures obtained are in very good agreement. We then performed many calculations of $\{\mathbf{h}_n^{\text{int}}\}$ to follow the development of magnetic order at lower temperature, captured the effect of multi-site local moment interactions for all systems from higher order local moment interactions, and produced phase diagrams.

The most striking results reported are for Mn$_3$A with cubic structure. Results of $\tilde{S}_{ss'}^{(2)}(\mathbf{q})$ obtained for cubic Mn$_3$Rh and Mn$_3$Ir are in stark contrast compared with those for Mn$_3$Pt. The theory directly links the instability of the collinear AFM phase to the pairwise interactions, which strongly favor the triangular AFM state in Mn$_3$Rh and Mn$_3$Ir and produce similar stability of both collinear and triangular AFM states in Mn$_3$Pt. Notably, we have provided the origin of the first-order phase transition between the triangular AFM and collinear AFM states in Mn$_3$Pt as a magnetovolume driven effect, and its absence in Mn$_3$Ir and Mn$_3$Rh. We constructed the temperature-pressure magnetic phase diagram of Mn$_3$Pt and obtained the same features as in experiment: compression destroys the collinear AFM order and eventually a single second-order PM-triangular AFM phase transition is observed after crossing a tricritical point. We also predict that the collinear AFM state is stable to low temperatures, hence ruling out the necessity to invoke other non-frustrated magnetic phases as suggested by Long [17]. In addition, by calculating the isothermal entropy change at the AFM-AFM transition and inspecting the magnetic phase diagram's features we showed that Mn$_3$Pt is not a good barocaloric material despite its strong first-order transition.

From this extensive study for all Mn$_3$A materials we have shown that the effect of multi-spin local moment interactions is not significant in tetragonal Mn$_3$A and so theoretical approaches based on effective pairwise interactions are sufficient.

However, for the cubic and hexagonal structures higher than quadratic order terms are non-negligible in the triangular AFM state, and so calculations going beyond a simple Heisenberg picture must be considered for an accurate description at finite temperature and different magnetic states. We also showed that leading magnetic interactions giving rise to the ferrimagnetic state in tetragonal Mn_3A are in overall satisfied and so this system should not be regarded as a frustrated magnetic material.

Finally, it should be stressed that SDFT-DLM theory can be used within a fully relativistic scheme so that SOC effects can be taken into account. Spin-dependent transport can also be modelled at different temperatures and states of magnetic order [44]. Although our results without SOC capture the major aspects of the magnetism of Mn_3A, the effect of the multi-site local moment interactions could be crucial to describe temperature-dependent properties and possible magnetic phase transitions when relativistic effects are taken into account, such as the weak FM component observed in the triangular AFM state for the hexagonal structure, and transitions between different chiral phases [1, 2, 27, 35]. This work lays out the groundwork for a future fully relativistic DFT-DLM investigation of Mn_3A.

References

1. Nagamiya T, Tomiyoshi S, Yamaguchi Y (1982) Triangular spin configuration and weak ferromagnetism of Mn_3Sn and Mn_3Ge. Solid State Commun. 42(5):385–388
2. Tomiyoshi S, Yamaguchi Y (1982) Magnetic structure and weak ferromagnetism of Mn_3Sn studied by polarized neutron diffraction. J Phys Soc Jpn 51(8):2478–2486
3. Krén E, Kádár G, Pál L, Sólyom J, Szabó P, Tarnóczi T (1968) Magnetic structures and exchange interactions in the mn-pt system. Phys Rev 171:574–585 Jul
4. Yasui H, Kaneko T, Yoshida H, Abe S, Kamigaki K, Mori N (1987) Pressure dependence of magnetic transition temperatures and lattice parameter in an antiferromagnetic ordered alloy Mn_3Pt. J Phys Soc Jpn 56(12):4532–4539
5. Krén E, Szabó P, Pál L, Tarnóczi T, Kádár G, Hargitai C (1968) Xray and susceptibility study of the firstorder magnetic transformation in Mn_3Pt. J Appl Phys 39(2):469–470
6. Krén E, Kádár G (1970) Neutron diffraction study of Mn_3Ga. Solid State Commun 8(20):1653–1655
7. Kádar G, Krén E (1971) Int J Magn 1:143
8. Kübler J, Felser C (2014) Non-collinear antiferromagnets and the anomalous hall effect. EPL (Europhys Lett) 108(6):67001
9. Qian JF, Nayak AK, Kreiner G, Schnelle W, Felser C (2014) Exchange bias up to room temperature in antiferromagnetic hexagonal Mn_3Ge. J Phys D: Appl Phys 47(30):305001
10. Sung NH, Ronning F, Thompson JD, Bauer ED (2018) Magnetic phase dependence of the anomalous hall effect in Mn_3Sn single crystals. Appl Phys Lett 112(13):132406
11. Zhang W, Han W, Yang S-H, Sun Y, Zhang Y, Yan B, Parkin SSP (2016) Giant facet-dependent spin-orbit torque and spin hall conductivity in the triangular antiferromagnet $IrMn_3$. Sci Adv 2:e1600759
12. Balke B, Fecher GH, Winterlik J, Felser C (2007) Mn_4Ga, a compensated ferrimagnet with high curie temperature and low magnetic moment for spin torque transfer applications. Appl Phys Lett 90(15):152504
13. Kurt H, Rode K, Venkatesan M, Stamenov P, Coey JMD (2011) High spin polarization in epitaxial films of ferrimagnetic Mn_3Ga. Phys Rev B 83:020405 Jan

14. Khmelevskyi S, Shick AB, Mohn P (2016) Prospect for tunneling anisotropic magneto-resistance in ferrimagnets: spin-orbit coupling effects in Mn$_3$Ge and Mn$_3$Ga. Appl Phys Lett 109(22):222402
15. Železný J, Zhang Y, Felser C, Yan B (2017) Spin-polarized current in noncollinear antiferro-magnets. Phys Rev Lett 119:187204 Nov
16. Zhang Y, Sun Y, Yang H, Železný J, Parkin SPP, Felser C, Yan B (2017) Strong anisotropic anomalous hall effect and spin hall effect in the chiral antiferromagnetic compounds Mn$_3$X(X = Ge, Sn, Ga, Ir, Rh, andPt). Phys Rev B 95:075128
17. Long MW (1991) A new magnetic structure for Mn$_3$Pt. J Phys: Conden Matter 3(36):7091
18. Khmelevskyi S, Ruban AV, Mohn P (2016) Magnetic ordering and exchange interactions in structural modifications of Mn$_3$Ga alloys: interplay of frustration, atomic order, and off-stoichiometry. Phys Rev B 93:184404
19. Tomeno I, Fuke HN, Iwasaki H, Sahashi M, Tsunoda Y (1999) Magnetic neutron scattering study of ordered Mn$_3$Ir. J Appl Phys 86(7):3853–3856
20. Krn E, Kdr G, Pl L, Slyom J, Szab P (1966) Magnetic structures and magnetic transformations in ordered Mn$_3$(Rh, Pt) alloys. Phys Lett 20(4):331–332
21. Shirai M, Tanimoto T, Motizuki K (1990) Successive magnetic phase transitions in Mn$_3$Pt. J Magn Magn Mater 90–91:157–158
22. Hirano Y, Suzuki N, Shirai M, Motizuki K (1995) Spin waves in Mn$_3$Pt. J Magn Magn Mater 140-144:1975–1976. International conference on magnetism
23. Kota Y, Tsuchiura H, Sakuma A (2008) Ab-initio study on the magnetic structures in the ordered Mn$_3$Pt. IEEE Trans Magn 44(11):3131
24. Zemen J, Mendive-Tapia E, Gercsi Z, Banerjee R, Staunton JB, Sandeman KG (2017) Frustrated magnetism and caloric effects in Mn-based antiperovskite nitrides: ab initio theory. Phys Rev B 95:184438 May
25. Boldrin D, Mendive-Tapia E, Zemen J, Staunton JB, Hansen T, Aznar A, Tamarit J-L, Barrio M, Lloveras P, Kim J, Moya X, Cohen LF (2018) Multisite exchange-enhanced barocaloric response in mn$_3$NiN. Phys Rev X 8:041035
26. Zimmer GJ, Krén E (1972) Investigation of the magnetic phase transformation in Mn$_3$Sn. AIP Conf Proc 5(1):513–516
27. Brown PJ, Nunez V, Tasset F, Forsyth JB, Radhakrishna P (1990) Determination of the magnetic structure of Mn$_3$Sn using generalized neutron polarization analysis. J Phys: Conden Matter 2(47):9409
28. Yamada N, Sakai H, Mori H, Ohoyama T (1988) Magnetic properties of -Mn$_3$Ge. Phys B+C 149(1):311 – 315
29. Nakatsuji S, Kiyohara N, Higo T (2015) Nature 527:202
30. Liu ZH, Zhang YJ, Liu GD, Ding B, Liu EK, Mehdi Jafri H, Hou ZP, Wang WH, Ma XQ, Wu GH (2017) Transition from anomalous hall effect to topological hall effect in hexagonal non-collinear magnet Mn$_3$Ga. Sci Rep 7:515
31. Nayak AK, Fischer JE, Sun Y, Yan B, Karel J, Komarek AC, Shekhar C, Kumar N, Schnelle W, Kübler J, Felser C, Parkin SSP (2016) Large anomalous hall effect driven by a nonvanishing berry curvature in the noncolinear antiferromagnet Mn$_3$Ge. Sci Adv 2(4)
32. Ohgushi K, Murakami S, Nagaosa N (2000) Spin anisotropy and quantum hall effect in the kagomé lattice: chiral spin state based on a ferromagnet. Phys Rev B 62:R6065–R6068 Sep
33. Taguchi Y, Oohara Y, Yoshizawa H, Nagaosa N, Tokura Y (2001) Science 291:2573
34. Chen H, Niu Q, MacDonald AH (2014) Anomalous hall effect arising from noncollinear antiferromagnetism. Phys Rev Lett 112:017205
35. Duan TF, Ren WJ, Liu WL, Li SJ, Liu W, Zhang ZD (2015) Magnetic anisotropy of single-crystalline Mn$_3$Sn in triangular and helix-phase states. Appl Phys Lett 107(8):082403
36. Sandratskii LM, Kübler J (1996) Role of orbital polarization in weak ferromagnetism. Phys Rev Lett 76:4963–4966 Jun
37. Winterlik J, Balke B, Fecher GH, Felser C, Alves MCM, Bernardi F, Morais J (2008) Structural, electronic, and magnetic properties of tetragonal mn$_{3-x}$Ga: experiments and first-principles calculations. Phys Rev B 77:054406

38. Zhang D, Yan B, Wu S-C, Kübler J, Kreiner G, Parkin SSP, Felser C (2013) First-principles study of the structural stability of cubic, tetragonal and hexagonal phases in Mn_3z (z = ga, sn and ge) heusler compounds. J Phys: Conden Matter 25(20):206006

39. Kurt H, Baadji N, Rode K, Venkatesan M, Stamenov P, Sanvito S, Coey JMD (2012) Magnetic and electronic properties of d022-Mn_3Ge (001) films. Appl Phys Lett 101(13):132410

40. Yang BS, Jiang LN, Chen WZ, Tang P, Zhang J, Zhang X-G, Yan Y, Han XF (2018) First-principles study of perpendicular magnetic anisotropy in ferrimagnetic d022-Mn_3x (x = ga, ge) on mgo and srtio3. Appl Phys Lett 112(14):142403

41. You Y, Guizhou X, Fang H, Gong Y, Liu E, Peng G, Feng X (2017) Designing magnetic compensated states in tetragonal Mn_3Ge-based heusler alloys. J Magn Magn Mater 429:40–44

42. Gutiérrez-Pérez RM, Antón RL, Załęski K, Holguín-Momaca JT, Espinosa-Magaña F, Olive-Méndez SF (2018) Tailoring magnetization and anisotropy of tetragonal mn3ga thin films by strain-induced growth and spin orbit coupling. Intermetallics 92:20–24

43. Bang H-W, Yoo W, Choi Y, You C-Y, Hong J-I, Dolinek J, Jung M-H (2016) Perpendicular magnetic anisotropy properties of tetragonal Mn_3Ga films under various deposition conditions. Current Appl Phys 16(1):63–67

44. Mankovsky S, Polesya S, Chadova K, Ebert H, Staunton JB, Gruenbaum T, Schoen MAW, Back CH, Chen XZ, Song C (2017) Temperature-dependent transport properties of FeRh. Phys Rev B 95:155139 Apr

Chapter 8
Summary and Outlook

In this thesis we have presented an ab initio theory for the calculation of pair- and multi-spin, or multi-site, magnetic interactions in magnetic materials. Our approach is able to describe the dependence of the electronic structure on the magnetic ordering associated with each atomic site, and in turn evaluates the effect that the electronic response has on the magnetism. This is obtained from modelling thermal fluctuations of magnetic degrees of freedom, which we describe by adopting the Disordered Local Moment picture. Local magnetic moments associated with atomic sites are pictured as emerging from the glue of many-interacting electrons and evolving on a time-scale long compared with the rest of electronic degrees of freedom. Within this view, Density Functional Theory calculations constrained to specific magnetic moment orientations are carried out to calculate a grand potential for the electrons and governing the local moments. The mathematical formalism of the method is suitable to deal with magnetic disorder through the construction of an effective medium set by the Coherent Potential Approximation. In this framework we have developed a methodology to describe first- and second-order magnetic phase transitions originated from the pairwise and multi-spin interactions, and hence depending on the electronic features too, and track them against temperature, magnetic field and atomic positions. A central step of our approach consists in the calculation of internal magnetic local fields that we used to extract the interactions among the local magnetic moments. In order to identify the most stable magnetic structures we firstly investigate the instabilities of the paramagnetic state by lattice Fourier transforming the magnetic interactions. Magnetic phase diagrams, and tricritical points in consequence, are directly obtainable from our approach. The method is also able to calculate isothermal entropy changes and adiabatic temperature changes, which we use to evaluate caloric effects at magnetic phase transitions and thus the refrigerating performance of a magnetic material.

The theory has been firstly applied to study the heavy rare earth elements in Chap. 5. We have found that most of their magnetism fundamentally originates from the mixed effect of both pairwise and multi-spin interactions. We have shown that the effect that the magnetic ordering of the f-electron local moments has on the valence electronic structure, common to all the heavy rare earths, establishes the origin of

© Springer Nature Switzerland AG 2020

E. Mendive Tapia, *Ab initio Theory of Magnetic Ordering*, Springer Theses,
https://doi.org/10.1007/978-3-030-37238-5_8

the general features among the magnetic phase transitions triggered by temperature and magnetic field. The qualitative changes induced on the Fermi surface topology by changing the magnetic ordering are demonstrated to be directly correlated to the stabilisation between a long-period helical antiferromagnetic structure and ferromagnetism at zero magnetic field. As a result, a generic temperature-magnetic field phase diagram for Gd to Ho has been produced, which describes first- and second-order phase transitions and predicts tricritical points in good agreement with experiment. The lanthanide contraction has been used to design a model based on Gd as a magnetic prototype and the de Gennes factor to scale the magnetic interactions.

In Chap. 6 we investigated the magnetic frustration in Mn-based antiperovskite Mn_3GaN system. The theory correctly predicts the transition temperature of a first-order phase transition between the paramagnetic state and an antiferromagnetic triangular structure, showing the linkage between the first-order character of the transition and the multi-spin interactions. We investigated the effect of biaxial strain on the magnetic frustration and predicted two new collinear magnetic phases at very high temperatures, which are a direct consequence of the strong dependence of the pairwise constants on the interatomic distances.

We have also carried out major studies comprising a very large range of collinear and non-collinear magnetic phases in cubic, hexagonal, and tetragonal lattices formed in Mn_3A (A=Sn, Ga, Ge, Pt, Ir, Rh) magnetic materials class, in Chap. 7. The theory describes magnetic phases and transition temperatures in good agreement with experiment. Albeit multi-spin interactions do not produce complicated magnetic phase transitions in these systems, we have found that their presence indicates that the calculation of pairwise interactions at different magnetic states would yet provide different values. Importantly, we have found that the first-order triangular antiferromagnetic to collinear antiferromagnetic transition in cubic Mn_3Pt is driven by magnetovolume sources and that multi-site effects are not necessary. The absence of this first-order magnetic phase transition for A=Ir and A=Rh has been also explained and attributed to a magnetovolume effect.

An investigation centred on caloric effects in the heavy rare earths, elastocaloric Mn_3GaN, and barocaloric Mn_3Pt, has been also performed. The magnetocaloric effect calculated for Gd and Dy is in qualitative and quantitative agreement with experiment. We have proposed a new elastocaloric cooling cycle showing both large temperature and entropy changes in Mn_3GaN, which exploits the magnetic phase diagram's features and the combination of first- and second-order magnetic phase transitions. We have also shown that Mn_3Pt is not a very good barocaloric material, despite its very complicated temperature-pressure magnetic phase diagram.

In summary, we have shown the validity and efficiency of SDFT-DLM theory in many different magnetic materials, providing explanations for many underlying physical phenomena behind magnetic phase transitions and useful predictions for refrigeration exploiting magnetism. Importantly, we have shown that our results are consequent of considering the mutual feedback between magnetic ordering and the electronic structure, and the connection of this to the presence of both pairwise and higher order multi-spin interactions.

We identify three different directions to further develop the theory and ideas presented in this thesis, with relevant importance in both fundamental and practical aspects:

1. The first refers to the usage of the pairwise and multi-spin interactions, that we show how to calculate from a pure first-principles approach, within atomistic modelling of magnetic systems. By means of atomistic spin model simulations, fundamental physics can be studied in detail from underlying microscopic origins [1], such as spin dynamics [2–4], spin torques [5] and anisotropy [6–8], exchange bias [9], and complex behaviour of magnetic nanoparticles [6], for example. The DFT basis of our approach motivates the linkage to atomistic models due to its broad use and applicability, as well as the natural treatment of atoms with associated local magnetic moments. In addition, our method directly comes from a robust formalism for the local moment thermal fluctuations, and hence it shows itself as a pertinent theory to be joined with atomistic schemes at finite temperatures. We have shown how the presence of multi-spin interactions has a profound impact on the magnetism as the magnetic ordering changes. However, only simpler pairwise constants are usually considered within atomistic calculations due to the lack of an appropriate ab initio theory for the calculation of more complicated exchange terms. Clearly, developing a method to adequately take into account the effect of multi-spin interactions from DFT-DLM theory into atomistic modellings has a great potential to be very useful and produce important results.

2. The second centres on studying further systems for spintronic applications. As shown in this thesis, the theory presents a very good performance for the description of antiferromagnetic states of different kind. It predicts the most potential antiferromagnetic states among all possible magnetic phase space, and tracks their temperature-dependent type and extent of long-range magnetic order. This gains, therefore, relevance in the context of antiferromagnetic spintronics, which is currently living an exciting period of time owing to recent novel advances, such as the demonstration of electrical switching and detection of Néel order [10, 11]. The prediction of spin-polarised current in triangular antiferromagnets from symmetry arguments [12], such as in Mn_3Sn studied in Chap. 7, also evidences the importance along this direction. Their frustrated magnetism is very similar to the originated in the antiperovskite structure, studied in Chap. 6, which also offers a broad chemical diversity. The theory can be used within a fully relativistic scheme so that spin-orbit coupling effects can be taken into account and spin-dependent transport can be modelled in different magnetic phases [13]. Our studies can be expanded, therefore, to address the mutual influence between currents and magnetic moment textures. Calculations centred on the effect of magnetic anisotropy on distortions and phase-coexistence observed in Mn_3A systems [14], as well as the fundamental role of chirality and Dzyaloshinskii–Moriya interaction on the family of Hall effects with both fully compensated and weekly ferromagnetic triangular states [15], can be performed too. Along these lines, the evaluation of new types of anisotropic high order correlations is extremely promising and

still unexplored. We believe that this together with our methodology to construct magnetic phase diagrams and monitoring the electronic properties against magnetic ordering could be very relevant for spintronic studies.

3. Finally, our theory is still lacking of a suitable description for the effect of lattice vibrations, which is of fundamental importance for the calculation of caloric effects at magnetic phase transitions. At the moment only a simple Debye model is used to obtain an adequate behaviour of the entropy against temperature, which is then used to evaluate adiabatic temperature changes. However, the atomic positions are fixed in our calculations. The direct consequence of this fact is that the phonon contribution to the caloric response at a first-order phase transition is not taken into account and that the mutual feedback between atomic displacements and magnetic degrees of freedom is neglected. We point out that thermal fluctuations of lattice displacements have been already studied using DFT technology in several works [16–18]. For example, by means of calculating effective forces a method to extract the impact of magnetic disorder on the phonon spectra has been achieved, obtaining remarkable results in the paramagnetic state of bcc iron [16]. However, in such an approach the CPA is circumvented in favour of a very efficient but less versatile description of thermal fluctuations. As a consequence, pairwise and multi-spin interactions, and their change due to the vibrational influence, are not addressed. Moreover, H. Ebert at al. have carried out theoretical progress using the CPA and assuming a displacement of the atomic potentials within a rigid muffin-tin approximation in the context of KKR formalism [17]. This work is a benchmark for our studies but further efforts are still necessary since the CPA was used solely to perform averages over atomic displacements and less rigorous supercell techniques were used to treat spin fluctuations. Moreover, a recent study of CrN has also shown that a dynamical coupling between magnetic fluctuations and lattice vibrations is necessary to explain finite-temperature phonon lifetimes, using a combined spin- and molecular dynamics approach [19]. Another method to tackle molecular dynamics in the disordered local moment picture has been also developed [20]. These works, among many others, have produced many important advances and represent cutting-edge theoretical techniques at present. However, they share in common that only zero temperature magnetic phases and/or fully disordered paramagnetism are adequately calculated. There is plenty of scope to develop our theory further along these advances, implementing our strategy to describe intermediate magnetic orders through the adequate consideration of pairwise and higher order interactions. Fundamental questions about entropy sources for caloric refrigeration and their relation to magneto-phonon coupling, tackling time and energy scale considerations, should be also addressed with more accuracy.

References

1. Evans RFL, Fan WJ, Chureemart P, Ostler TA, Ellis MOA, Chantrell RW (2014) Atomistic spin model simulations of magnetic nanomaterials. J Phys Condens Matter 26(10):103202
2. Ostler TA, Evans RFL, Chantrell RW, Atxitia U, Chubykalo-Fesenko O, Radu I, Abrudan R, Radu F, Tsukamoto A, Itoh A, Kirilyuk A, Rasing T, Kimel A (2011) Crystallographically amorphous ferrimagnetic alloys: comparing a localized atomistic spin model with experiments. Phys Rev B 84:024407
3. Radu I, Vahaplar K, Stamm C, Kachel T, Pontius N, Dürr HA, Ostler TA, Barker J, Evans RFL, Chantrell RW, Tsukamoto A, Itoh A, Kirilyuk A, Rasing T, Kimel AV (2011) Transient ferromagnetic-like state mediating ultrafast reversal of antiferromagnetically coupled spins. Nature 472:205
4. Dupé B, Bihlmayer G, Böttcher M, Blügel S, Heinze S (2016) Engineering skyrmions in transition-metal multilayers for spintronics. Nat Commun 7:11779
5. Chureemart P, Evans RFL, Chantrell RW (2011) Dynamics of domain wall driven by spin-transfer torque. Phys Rev B 83:184416
6. Garanin DA, Kachkachi H (2003) Surface contribution to the anisotropy of magnetic nanoparticles. Phys Rev Lett 90:065504
7. Asselin P, Evans RFL, Barker J, Chantrell RW, Yanes R, Chubykalo-Fesenko O, Hinzke D, Nowak U (2010) Constrained Monte Carlo method and calculation of the temperature dependence of magnetic anisotropy. Phys Rev B 82:054415
8. Gino H, Woodcock TG, Colin F, Alexander G, Julian D, Thomas S, Oliver G (2010) The role of local anisotropy profiles at grain boundaries on the coercivity of $Nd_2Fe_{14}B$ magnets. Appl Phys Lett 97(23):232511
9. Mitsumata C, Sakuma A, Fukamichi K (2003) Mechanism of the exchange-bias field in ferromagnetic and antiferromagnetic bilayers. Phys Rev B 68:014437
10. Wadley P, Howells B, Železný J, Andrews C, Hills V, Campion RP, Novák V, Olejník K, Maccherozzi F, Dhesi SS, Martin SY, Wagner T, Wunderlich J, Freimuth F, Mokrousov Y, Kuneš J, Chauhan JS, Grzybowski MJ, Rushforth AW, Edmonds KW, Gallagher BL, Jungwirth T (2016) Electrical switching of an antiferromagnet. Science 351(6273):587–590
11. Železný J, Gao H, Výborný K, Zemen J, Mašek J, Manchon A, Wunderlich J, Sinova J, Jungwirth T (2014) Relativistic Néel-order fields induced by electrical current in antiferromagnets. Phys Rev Lett 113:157201
12. Jakub Ž, Yang Z, Claudia F, Binghai Y (2017) Spin-polarized current in noncollinear antiferromagnets. Phys Rev Lett 119:187204
13. Mankovsky S, Polesya S, Chadova K, Ebert H, Staunton JB, Gruenbaum T, Schoen MAW, Back CH, Chen XZ, Song C (2017) Temperature-dependent transport properties of FeRh. Phys Rev B 95:155139
14. Duan TF, Ren WJ, Liu WL, Li SJ, Liu W, Zhang ZD (2015) Magnetic anisotropy of single-crystalline Mn_3Sn in triangular and helix-phase states. Appl Phys Lett 107(8):082403
15. Zhang Y, Sun Y, Yang H, Železný J, Parkin SPP, Felser C, Yan B (2017) Strong anisotropic anomalous hall effect and spin hall effect in the chiral antiferromagnetic compounds Mn_3X (X = Ge, Sn, Ga, Ir, Rh, and Pt). Phys Rev B 95:075128
16. Körmann F, Dick A, Grabowski B, Hickel T, Neugebauer J (2012) Atomic forces at finite magnetic temperatures: phonons in paramagnetic iron. Phys Rev B 85:125104
17. Ebert H, Mankovsky S, Chadova K, Polesya S, Minár J, Ködderitzsch D (2015) Calculating linear-response functions for finite temperatures on the basis of the alloy analogy model. Phys Rev B 91:165132
18. Shun-Li S, Yi W, Zi-Kui L (2010) Thermodynamic fluctuations between magnetic states from first-principles phonon calculations: the case of bcc Fe. Phys Rev B 82:014425

19. Irina S, Anders B, Albert G, Tilmann H, Fritz K, Blazej G, Jörg N, Björn A (2018) Anomalous phonon lifetime shortening in paramagnetic crn caused by spin-lattice coupling: a combined spin and ab initio molecular dynamics study. Phys Rev Lett 121:125902
20. Peter S, Björn A, Abrikosov IA (2012) Equation of state of paramagnetic crn from ab initio molecular dynamics. Phys Rev B 85:144404

Appendix
The CPA Equations in the Paramagnetic Limit

In this section we provide detail on the derivation of Eq. (3.100). To arrive to this result we solve the CPA equations of disordered local moments, given in Eq. (3.40), in the paramagnetic limit, i.e. when approaching the fully disordered state $\{\mathbf{m}_n\} = \{\mathbf{0}\}$. In terms of the impurity matrix, the paramagnetic state is described to the lowest order in $\{\mathbf{m}_n\}$ as

$$\underline{D}_i(\hat{e}_i) = \underline{D}_i^0 + \delta\underline{D}_i(\hat{e}_i), \tag{A.1}$$

where $\underline{D}_i^0$ is given by Eq. (3.82) (or equivalently by Eq. (3.84)) and

$$\delta\underline{D}_i(\hat{e}_i) = \underline{D}_i^0\left[\delta\underline{\Delta}_i^0\underline{\tau}_{c\text{-}0,ii} + \underline{\Delta}_i^0\delta\underline{\tau}_{c,ii}\boldsymbol{\sigma}\cdot\hat{m}_i\right]\underline{D}_i^0, \tag{A.2}$$

as can be seen from Eq. (3.90). Hence, the CPA equations are expressed as

$$\int d\hat{e}_i\, P_i^0 \underline{D}_i^0 = \underline{I}_2, \tag{A.3}$$

for the zeroth order and

$$\int d\hat{e}_i\, \left(\delta P_i(\hat{e}_i)\underline{D}_i^0 + P_i^0\delta\underline{D}_i(\hat{e}_i)\right) = (0), \tag{A.4}$$

for the first order. Since $\int d\hat{e}_i\, P_i^0 \boldsymbol{\sigma}\cdot\hat{e}_i = 0$, it directly follows from the zeroth order that

$$\frac{1}{2}\left(\underline{D}_{i+}^0 + \underline{D}_{i-}^0\right) = 1. \tag{A.5}$$

By performing some simple manipulations into Eq. (A.5) one can show that

$$\frac{1}{2}\left(\underline{t}_{i+}^{-1} + \underline{t}_{i+}^{-1}\right) - \underline{t}_{c\text{-}0,i}^{-1} = -\left(\underline{t}_{i+}^{-1} - \underline{t}_{c\text{-}0,i}^{-1}\right)\underline{\tau}_{c\text{-}0,ii}\left(\underline{t}_{i+}^{-1} - \underline{t}_{c\text{-}0,i}^{-1}\right). \tag{A.6}$$

© Springer Nature Switzerland AG 2020

E. Mendive Tapia, *Ab initio Theory of Magnetic Ordering*, Springer Theses,

https://doi.org/10.1007/978-3-030-37238-5

This is a central equation for the effective medium quantities $\underline{t}_{c\text{-}0,i}^{-1}$ and $\underline{\tau}_{c\text{-}0,ii}$ in terms of $\underline{t}_{i+}^{-1}$ and $\underline{t}_{i-}^{-1}$. In order to develop the first order we split Eq. (A.4) into three different terms,

$$\frac{3}{4\pi} \int d\hat{e}_i \, \mathbf{m}_i \cdot \hat{e}_i \underline{D}_i^0 , \tag{A.7}$$

$$\frac{1}{4\pi} \int d\hat{e}_i \, \underline{D}_i^0 \left(-\delta \underline{t}_{c,i}^{-1} \boldsymbol{\sigma} \cdot \hat{m}_i \right) \underline{D}_i^0 , \tag{A.8}$$

$$\frac{1}{4\pi} \int d\hat{e}_i \, \underline{D}_i^0 \left(\frac{1}{2} \left(\underline{t}_{i+}^{-1} + \underline{t}_{i-}^{-1} \right) \underline{L}_2 + \frac{1}{2} \left(\underline{t}_{i+}^{-1} - \underline{t}_{i-}^{-1} \right) \boldsymbol{\sigma} \cdot \hat{e}_i - \underline{t}_{c\text{-}0,i}^{-1} \underline{L}_2 \right) \delta \underline{\tau}_{c,ii} \boldsymbol{\sigma} \cdot \hat{m}_i \underline{D}_i^0 , \tag{A.9}$$

where we have used definitions given in Eqs. (3.59), (3.91), and (3.92). The way to proceed now is to exploit Eq. (A.5) in order to get rid of the terms $(\underline{D}_{i+}^0 + \underline{D}_{i-}^0)/2$ inside $\underline{D}_i^0$ and multiply by $(\underline{D}_{i+}^0)^{-1}$ and $(\underline{D}_{i-}^0)^{-1}$, in the left and right hand sides respectively, to deal with $(\underline{D}_{i+}^0 - \underline{D}_{i-}^0)/2$. To simplify further Eq. (A.8) it is also necessary to use Eq. (A.6) as well as the fact that

$$\frac{1}{2} \left(\underline{D}_{i+}^0 - \underline{D}_{i-}^0 \right) = 1 - \underline{D}_{i-}^0 = \underline{D}_{i+}^0 - 1 , \tag{A.10}$$

which follows from Eq. (A.5). Note that only terms accompanied by $(\boldsymbol{\sigma} \cdot \hat{m}_i)$, $(\mathbf{m}_i \cdot \hat{e}_i)(\boldsymbol{\sigma} \cdot \hat{e}_i)$, or $(\boldsymbol{\sigma} \cdot \hat{e}_i)(\boldsymbol{\sigma} \cdot \hat{m}_i)(\boldsymbol{\sigma} \cdot \hat{e}_i)$ give non-zero contributions after performing the integral and that

$$\frac{3}{4\pi} \int d\hat{e}_i \, (\mathbf{m}_i \cdot \hat{e}_i)(\boldsymbol{\sigma} \cdot \hat{e}_i) = m_i \sigma_z , \tag{A.11}$$

$$\frac{1}{4\pi} \int d\hat{e}_i \, (\boldsymbol{\sigma} \cdot \hat{m}_i) = \frac{1}{4\pi} \int d\hat{e}_i \, (\boldsymbol{\sigma} \cdot \hat{e}_i)(\boldsymbol{\sigma} \cdot \hat{m}_i)(\boldsymbol{\sigma} \cdot \hat{e}_i) = \sigma_z . \tag{A.12}$$

After carrying out this lengthy algebra one obtains

$$\frac{1}{2} m_i \left(\underline{t}_{i+}^{-1} - \underline{t}_{i-}^{-1} \right) - \left(\underline{t}_{i+}^{-1} - \underline{t}_{c\text{-}0}^{-1} \right) \delta \underline{\tau}_{c,ii} \left(\underline{t}_{i-}^{-1} - \underline{t}_{c\text{-}0}^{-1} \right)$$
$$+ \delta \underline{t}_{c,i}^{-1} + \left(\underline{t}_{i+}^{-1} - \underline{t}_{c\text{-}0}^{-1} \right) \underline{\tau}_{c\text{-}0,ii} \delta \underline{t}_{c,i}^{-1} + \delta \underline{t}_{c,i}^{-1} \underline{\tau}_{c\text{-}0,ii} \left(\underline{t}_{i-}^{-1} - \underline{t}_{c\text{-}0}^{-1} \right) = 0 , \tag{A.13}$$

where the first and second terms originate from Eqs. (A.7) and (A.9), respectively, and the rest from Eq. (A.8). We can now rearrange the terms of this equation by taking into account that $\delta \underline{t}_{c,i}^{-1}$ and $\delta \underline{\tau}_{c,ii}$ are related by

$$\delta \underline{\tau}_{c,ii} = - \sum_k \underline{\tau}_{c\text{-}0,ik} \delta \underline{t}_{c,k}^{-1} \underline{\tau}_{c\text{-}0,ki} \tag{A.14}$$

which follows from Eq. (3.96), so that

$$\left[1 + \left(\underline{t}_{i+}^{-1} - \underline{t}_{c\text{-}0}^{-1} \right) \underline{\tau}_{c\text{-}0,ii} \right] \delta \underline{t}_{c,i}^{-1} \left[1 + \underline{\tau}_{c\text{-}0,ii} \left(\underline{t}_{i-}^{-1} - \underline{t}_{c\text{-}0}^{-1} \right) \right]$$
$$= -\frac{1}{2} m_i \left(\underline{t}_{i+}^{-1} - \underline{t}_{i-}^{-1} \right) - \left(\underline{t}_{i+}^{-1} - \underline{t}_{c\text{-}0}^{-1} \right) \left[\sum_{k \neq i} \underline{\tau}_{c\text{-}0,ik} \delta \underline{t}_{c,k}^{-1} \underline{\tau}_{c\text{-}0,ki} \right] \left(\underline{t}_{i-}^{-1} - \underline{t}_{c\text{-}0}^{-1} \right) . \tag{A.15}$$

From this result we can write an equation for the derivative of $\underline{t}_{c,i}^{-1}$ with respect to an arbitrary order parameter $\mathbf{m}_j$. We firstly multiply by $\underline{D}_{i+}^0$ and $\underline{D}_{i-}^0$ at the left and right hand sides of Eq. (A.15), respectively. After this we divide the equation by $\delta\mathbf{m}_j$ and take the paramagnetic limit, which gives

$$\underline{\Lambda}_{ij} \equiv \frac{\partial \underline{t}_{c,i}^{-1}}{\partial \mathbf{m}_j} = \frac{1}{2}\left(\underline{X}_{i-}^0 - \underline{X}_{i+}^0\right)\delta_{ij} - \underline{X}_{i+}^0\left[\sum_{k\neq i}\underline{\tau}_{c\text{-}0,ik}\frac{\partial \underline{t}_{c,k}^{-1}}{\partial \mathbf{m}_j}\underline{\tau}_{c\text{-}0,ki}\right]\underline{X}_{i-}^0,\quad (A.16)$$

where $\underline{X}_{i+}^0$ and $\underline{X}_{i-}^0$ are the components of the zeroth order of the excess scattering matrix in the spin frame of reference in which it is diagonal,

$$\underline{X}_i^0 = \begin{pmatrix} X_{i+}^0 & \underline{0} \\ \underline{0} & X_{i-}^0 \end{pmatrix} = \begin{pmatrix} (\underline{t}_{i+}^{-1} - \underline{t}_{c\text{-}0,i}^{-1}) + \underline{\tau}_{c\text{-}0,ii} & \underline{0} \\ \underline{0} & (\underline{t}_{i-}^{-1} - \underline{t}_{c\text{-}0,i}^{-1}) + \underline{\tau}_{c\text{-}0,ii} \end{pmatrix}^{-1}. \quad (A.17)$$

The next step is to apply the lattice Fourier transform introduced in Eq. (3.68), that is

$$\tilde{\underline{\Lambda}}_{ss'}(\mathbf{q}) = \frac{1}{N_c}\sum_{tt'}\underline{\Lambda}_{ts\,t's'}e^{i\mathbf{q}\cdot(\mathbf{R}_t+\mathbf{r}_s-\mathbf{R}_{t'}-\mathbf{r}_{s'})}. \quad (A.18)$$

Equation (A.16) then becomes

$$\tilde{\underline{\Lambda}}_{ss'}(\mathbf{q}) - \underline{X}_{s+}^0\underline{\tau}_{c\text{-}0,ss}\tilde{\underline{\Lambda}}_{ss}(\mathbf{q})\underline{\tau}_{c\text{-}0,ss}\underline{X}_{s-}^0 - \frac{1}{2}\left(\underline{X}_{s-}^0 - \underline{X}_{s+}^0\right)$$

$$= -\frac{1}{N_c}\sum_{tt'}e^{i\mathbf{q}\cdot(\mathbf{R}_t+\mathbf{r}_s-\mathbf{R}_{t'}-\mathbf{r}_{s'})}\underline{X}_{s+}^0\left(\sum_{t''s''}\underline{\tau}_{c\text{-}0,ts,t''s''}\underline{\Lambda}_{t''s'',t's'}\underline{\tau}_{c\text{-}0,t''s'',ts}\right)\underline{X}_{s-}^0$$

$$\text{(A.19)}$$

$$= -\frac{1}{N_c}\sum_{tt'}e^{i\mathbf{q}\cdot(\mathbf{R}_t+\mathbf{r}_s-\mathbf{R}_{t'}-\mathbf{r}_{s'})}\underline{X}_{s+}^0\left(\sum_{t''s''}\underline{\tau}_{c\text{-}0,ts,t''s''}\underline{\Lambda}_{t''s'',t's'}\right.$$

$$\times\left.\frac{1}{V_{\text{BZ}}}\int_{V_{\text{BZ}}}\mathrm{d}\mathbf{k}\,\underline{\tau}_{c\text{-}0,s'',s}(\mathbf{k})e^{-i\mathbf{k}\cdot(\mathbf{R}_{t''}+\mathbf{r}_{s''}-\mathbf{R}_t-\mathbf{r}_s)}\right)\underline{X}_{s-}^0,$$

where V_{BZ} is the volume of the Brillouin zone. Conveniently rearranging the exponential factors and performing the sums invoking the convolution theorem finally gives

$$\tilde{\underline{\Lambda}}_{ss'}(\mathbf{q}) = \frac{1}{2}\left(\underline{X}_{s-}^0 - \underline{X}_{s+}^0\right)$$

$$- \underline{X}_{s+}^0\left[\sum_{s''}\frac{1}{V_{\text{BZ}}}\int_{V_{\text{BZ}}}\mathrm{d}\mathbf{k}\,\tilde{\underline{\tau}}_{c\text{-}0,ss''}(\mathbf{k}+\mathbf{q})\tilde{\underline{\Lambda}}_{s''s'}(\mathbf{q})\tilde{\underline{\tau}}_{c\text{-}0,s''s}(\mathbf{k}) - \underline{\tau}_{c\text{-}0,ss}\tilde{\underline{\Lambda}}_{ss}(\mathbf{q})\underline{\tau}_{c\text{-}0,ss}\right]\underline{X}_{s-}^0$$

$$\text{(A.20)}$$

The manufacturer's authorised representative in the EU is Springer
Nature Customer Service Centre GmbH, Europaplatz 3, 69115 Heidelberg,
Germany. If you have any concerns regarding our products, please
contact ProductSafety@springernature.com

Printed and bound by CPI Group (UK) Ltd, Croydon, CR0 4YY
28/11/2025
02007724-0007